The Dimensional Theory of Everything

William H. Johnson III

The Dimensional Theory of Everything

Copyright © 2022 William H. Johnson III

All rights reserved.

ISBN: 9798780966913

DEDICATION

First, and foremost, I dedicate this book to Jesus, who is the author and finisher of my faith; to God be the glory forever and ever. I dedicate this book to my wife and family for loving me as I am. I dedicate this book to everyone who believes on the name of Jesus; you were the hands and feet of him who brought me to this place. I dedicate this book to everyone who has personally known me. I have no doubt that you had to put up with me in some way. Please forgive me for my many failures. Some of you will recognize yourself in here. I did not use anyone's name out of respect. May you be protected from the mockery to be cast upon me. I love you all.

CONTENTS

	Acknowledgments	i
1	EVERYTHING THAT YOU CAN NAME BELONGS IN A DIMENSION	1
2	THE THEORY OF EVERYTHING	1
3	WHY PUBLISH NOW?	2
4	MY MODEL ALLOWS ME TO GAIN PERSPECTIVE ON ANY TOPIC	3
5	HAVE I LOST MY MIND?	3
6	THE PROOF IS IN THE PUDDING	4
7	THE BUILDING BLOCKS OF LOGIC	5
8	MRS. CROOK, MY SECOND GRADE TEACHER	6
9	MR. JORDAN, MY GEOMETRY TEACHER	7
10	EINSTEIN'S WORLD	8
11	DEFINE DIMENSION IN THE SPACE-TIME CONTINUUM	8
12	CHALLENGE TO THE SCIENTIFIC COMMUNITY	10
13	REALITY IS BIGGER THAN THE UNIVERSE	12
14	HOW MANY DIMENSIONS DO YOU FUNCTION IN?	12
15	YOUR UNSEEN PARTS	13
16	THE DIMENSIONS BEYOND TIME	16
17	IN HIS IMAGE	18
18	PAUL'S PRAYER HITS THE HIGH POINTS	18
19	THE BIG BANG BIRTHED THE PHYSICAL UNIVERSE	19
20	THE LAMB'S BOOK OF LIFE	21
21	THE TWELFTH DIMENSION	22
22	DIMENSIONAL CONTINUUM FROM OUR PERSPECTIVE	23
23	DIMENSIONAL CONTINUUM FROM HIS PERSPECTIVE	23

24	THE LAST DIMENSION	24
25	THE BIG BANG IN THE UNSEEN UNIVERSE	24
26	WHAT WE DO KNOW	25
27	FROM HEAVEN TO EARTH	25
28	THE LAKE OF FIRE	26
29	THE BURNING BUSH	26
30	THE RETURN OF JESUS	28
31	NEW HEAVEN AND NEW EARTH	28
32	THE NEW JERUSALEM	29
33	YOUR IMMORTAL SOUL	31
34	WORDS ARE SPIRIT	31
35	MY PERSONAL HISTORY	32
36	SALVATION	32
37	128	33
38	COLLEGE	34
39	JANUARY 28, 1986 SPACE SHUTTLE CHALLENGER DISASTER	35
40	EUREKA!	35
41	DIMENSIONS	36
42	THE STAR OF BETHLEHEM	36
43	SOMEONE WALKED THROUGH THE WALL	37
44	INSTANTANEOUS PHYSICAL HEALING	39
45	THE RUSH OF A MIGHTY WIND	39
46	GOD SPOKE TO ME, ALOUD, I HEARD HIM WITH MY EARS	40
47	THE VOICE OF GOD	41
48	FEAR THE LORD	41
49	YOU TOO	42

50	SUCH A TIME AS THIS	44
51	SUPERNATURAL EXPERIENCES	45
52	THE SANCTUARY	47
53	THE TRINITY	50
54	QUOTES FROM ALBERT EINSTEIN	51
55	ODE TO STAR TREK	52
56	LIFE IN YOUR DREAMS	53
57	PROPHECY	55
58	AN INCONVENIENT TRUTH	55
59	LIFE IS ALL ABOUT THE HEART	57
60	HIDDEN THINGS	59
61	THE END FROM THE BEGINNING	60
62	THE DIVIDING LINE	60
63	ONCE SAVED ALWAYS SAVED	60
64	WITH WHOM DID GOD MAKE COVENANTS?	61
65	HOT OFF THE PRESS	63
66	BELIEF AND UNBELIEF	64
67	ONE GOD	67
68	DAYS OF CREATION	68
69	SCIENCE SAYS	.70
70	RELATIONSHIP IS EVERYTHING	73
71	BELIEVE	73
72	HARDENING OF THE HEART	74
73	SUMMARY	76
74	PRACTICALLY SPEAKING	82
75	CHALLENGE TO THE WHOLE WORLD	83

76	NOTES TO SELF	84
77	THE END?	85
78	BE SAVED AND AVOID THE SECOND DEATH	85
79	CONCLUSION	86
80	TALK TO GOD THROUGH YOUR INTERCESSOR, JESUS	86
81	GO FORTH AND BE BAPTIZED	87

ACKNOWLEDGMENTS

I would like to express my special thanks and gratitude to everyone who has taught me anything. I'm not going to call you by name, to protect the innocent. However, two of my public school teachers were particularly instrumental in showing me who I am and I want to thank them individually. Mrs. Crook, my second grade teacher, who gave the very first spark that made learning fun. Mr. Jordan, my tenth grade geometry teacher, who gently pulled me in the direction of math, science, and reason to set me on my path. I want to give my heartfelt thanks to you all.

1 Everything that You Can Name Belongs in a Dimension

The dimensional theory of everything has one overriding precept; everything that you can name belongs in a dimension, or in multiple dimensions, and all of the dimensions, which define all of reality, form a continuum connected to the space-time continuum that Einstein gave us. Einstein got it right, as far as he went, but he and the whole scientific community have been looking for the remainder of the theory in the wrong direction ever since. I'm here to set the scientific world on the right path, so that they can do what they're good at, using math, observations, measurements, and experiments to put flesh and bone on this theory. More than that, I'm here to expand your understanding of reality and give you a model that you can use to gain perspective on everything, every day.

2 The Theory of Everything

A theory of everything is a hypothesis used to explain everything that exists. The scientific community has searched for such a theory of everything ever since Albert Einstein's theory of relativity sparked the imagination of scientists all over the world in the early 1900's. A theory is just that, it's a theory; it's not a scientifically proven model. Einstein's theory eventually became increasingly proven through experiment, observation and measurement; it predicted specific implications, many of which have since been observed in nature. Einstein spent his later years trying to find the theory of everything, but he failed, and no one has been successful, until now.

In the best case scenario, such a theory should provide insight that allows us to predict detailed outcomes and trends, which make it easier to go looking for them, to prove or disprove the model. At the very least, a good theory can reduce time wasted in unfruitful directions. However, the bar for the correct theory of everything must be much higher. It must provide so many insights that the whole world continually makes more sense, all fields of study should be able to advance rapidly because they are looking in the right direction.

This is exactly what I have been doing with the dimensional theory of everything for nearly four decades. I have spent a lifetime watching documentaries about all manner of academic pursuits in an attempt to accomplish one of two things; either I would learn just enough about the topic to see the future of the field, or I would make predictions about the field that would later prove to be incorrect. Astonishingly, I have observed countless predictions that were correct and none that were proven incorrect; everything I have paid attention to, and applied the insight from this theory upon, and had the good fortune to see the field advance, has become the confirmation that I needed to bring you this theory now.

3 Why Publish Now?

I'm currently 62 years old, today being December 7, 2021, 80 years to the day from Pearl Harbor, that day that will go down in infamy, and now we're going into 2022, at the time of writing. I have no idea why I do that, but I relate everything to everything, it seems to be my slant; for nothing is more important than relationship. I back that statement up with logic, later in this book.

If I am going to present a theory of everything, clearly I need to know that nothing exists, which I can recognize, that does not fit into the theory. That requirement sent me on a life-long journey to disprove my own theory but it's really difficult to prove something like that, the contrary, that is. I do not claim to know everything, but it is no small inconvenience to my wife, and likely others, that it appears that I think I have to know everything.

I've tried to hide it, but those close to me know that I cannot resist an investigation. To them, I apologize for being such a pain in the rear end. Thank you for loving me anyway.

So, I have been battling COVID for the past month, and it tried to kill me, but clearly I'm too ornery for that to happen, just yet. I ain't no spring chicken, and tomorrow is not promised to anyone. My biggest take-away is this, before I die, it is my responsibility to give this theory to the world, so **here I am**. I'm prepared to suffer the ridicule that will undoubtedly come my way.

4 My Model Allows Me to Gain Perspective on Any Topic

I believe that if you tell me enough about any topic, so that I understand where you are in your journey, and the problems that you are facing, I can gain perspective on anything. I might not provide you with perspective that contains the specifics that you desire, but I do it over and over, I've done it so many times that I failed to keep up with them all. In the beginning, I wrote on every scrap of paper in sight, many of which I collected, but having moved my physical belongings from house to house, over the decades, and I find that I have done a poor job of keeping all the notes. I have boxes full of notes, but I have not opened them since I packed them. I see the whole world through this lens, it is my reality, and I don't need notes.

It really doesn't matter that I have lost some notes because I can make the same discoveries all over again, but the disappointing part is that those notes showed what I knew and when I knew it, to some degree, and that, in itself has some measureable quality to it because we can see where we are now, relative to the predictions recorded in them. I'm sure there's plenty of that in those boxes.

5 Have I Lost My Mind?

I can tell you that the eureka-moment was fascinating moment, in itself. It flushed my face and my mind, well, exploded. I had told people for years before I had the theory that I was going to solve this mystery, and by the grace of god, I did arrive at the answer, but I have to tell you that I'm not really that smart, I didn't figure it out, it was given to me.

While I have discussed various parts with various people, over the years, I basically kept my trap shut. I guess you could describe me as a closet theorist, initially because even I thought I was crazy, and a substantial effort was spent examining myself. I began to expose astounding results and it sounded like I was having delusions of grandeur, to myself. I was so shocked that I was asking myself, what manner of thing is this? The model produces results in the most unexpected ways, it just works, and the more it becomes your way of thinking, the greater the perspective gain.

That quickly became the hallmark of the theory; I can gain perspective on anything. Who in the world would allow a statement like that out of their mouth? Either I'm onto something or I'm mentally deranged, I don't know how else to describe it. That statement is child's play. I intend to gain perspective on the big bang in this book, and perhaps so many other topics that you should have a need to pick your jaw up off the floor. In fact, I'm going to explain the big bang more completely and cohesively than anyone before me because my model of everything is better than any model before it.

I intend to give you more astounding statements than you have ever read in one book, and I won't even be trying hard. Listen to that puffed up language, I can't stand it, but it is my reality, all day, every day, this is continually going on inside me. I might as well warn you, I'm going to step on some toes in this book.

I hope you can set early judgment of the messenger aside, and not count it as arrogance, because it is not. Even so, I'm going to kill the multiverse theory, forever, and others, right here in this book. See, I can't help it, how do you say that and not sound puffed up? The proof is in the pudding, that's the only way I can respond.

Eventually, you seriously have to put me into one of those camps, for yourself, and I had to do it for myself, as well. Am I a nut job, or aren't I? That question was thankfully resolved and the concern subsided in me after a couple of years, but I didn't reach a fast conclusion, I needed to know, and I didn't shut the door on the possibility of either outcome for quite some time. I'm just relieved that it settled in a positive direction because it was terribly unsettling at the time.

6 The Proof is in the Pudding

If the model didn't produce fruitful results, then it would be useless. If it is true, then is must be useful; that was my benchmark. I also needed to be able to explain it to anyone; a theory of everything must pull things together to become simplified, not increase the complexity.

Many of you will be disappointed because it doesn't provide you with $E=MC2$. However, it absolutely can lead to those equations by pointing you in the direction in which you should look for them. That is not my forte. I am not a mathematician or even a scientist. To be honest, I don't even have an interest in it. I would be open to working with some of you on your topics, with hopes that rapid perspective gain could assist in moving your exploration along. By the same token, some things simply don't interest me. I do look at all manner of things, and I'd try to be open, but the reality is you don't need me.

You can understand the theory and apply it for yourself. It might be slower at first, but it pays its own dividends. It just works. I proved it for myself hundreds of times, over and over again. It's useful or it's not; if there's truth in it, it must be useful.

If you want my assistance, I need you to boil your topic down to a two-hour presentation, and it needs to have the main points you believe, the evidence that you have to support what you believe, and the problems you can't solve. That's it, and I can understand from only that, which parts are on track for advancement, and which parts are spinning your wheels. Many times, the perspective gain is far more specific than that. Sometimes it isn't, and I really have no control over that part. More information than a two-hour presentation is better, but that is the lowest bar that must be reached if I am to be of any significant use to you.

I'm not a mind reader or a fortune teller or god. I cannot see the future, well, I suppose it depends on how you define that, but I do not claim any special powers, at all. Everything I do, you can do, and you can likely do it better than me in the long run. I'm not asking god to solve your problem, and I certain am not solving your problem. However, I can gain fruitful perspective that will assist you in finding your answer, because the model works, it puts everything in its place, and that is perspective gained.

The very definition of perspective gain is the definition of moving from one dimension to another. Ah, definitions, we need a common language. No sense in dilly-dallying, let's take a peek.

7 The Building Blocks of Logic

The definition of a dimension is unique perspective. Look at the period at the end of that statement, there is no ifs, ands, or buts. A dimension cannot be defined any more succinctly or accurately. If any man can define a dimension better than that, then my theory is incorrect.

The first corollary is this: in order to gain perspective on one dimension, one must observe that dimension from another dimension. The term observe obviously has its concrete definition in the physical dimensions that make up three-dimensional space, but we extend the concept into other dimensions conceptually, and we do that by using the first part of the statement, which is by gaining perspective. If the solution forms a continuum, then the same corollary must continue to apply throughout the whole of the continuum.

In fact, that statement is the key to the proof. We will see how it applies to the physical dimensions, and then we will use it on the upper dimensions. The second corollary is this, if the definition of a dimension is unique perspective, and then there can only be one of them in existence. If there is no unique perspective, it is not a dimension. If something fits in one dimension, you don't get to make another dimension with the same purpose.

Now, that last statement, on the surface, might seem to be a bit ambiguously stated for this reason, for example, a three dimensional object is physically located in multiple dimensions at the same time, it sits in four dimensions, to be exact. All physical objects sit in three-dimensional space at some point in time; that defines the same four-dimensional reality for any and all inanimate physical objects. Einstein defined these four dimensions as a continuum for us, and I added the clarification that no physical object can exist anywhere else without breaking the definition of a dimension.

You cannot have physical height and width and depth anywhere, by any theory, or any mathematical possibility, or any imagination, and have any logical integrity. The only way a physical multiverse theory can hold water, is for that theory to define a dimension in a different way. Unique is unique, are you going to give that word a new definition, too? Dimension has already been proven by Einstein to be the central structure of reality, so what word do you plan to use when you redefine Einstein's theory? I digress.

Back to my ambiguous example, one might argue that the three dimensions have the same purpose, so you can see how the description of unique purpose is a bit skewed, but unique purpose is not the dividing line, the dividing line is perspective gain from dimension to dimension. In a moment, we're going to take a quick walk through the lower dimensions to see that corollary more clearly, but first, let's get our language together, so we understand one another.

Let's agree on what a corollary is. The dictionary says it's an easily drawn conclusion. I use the word corollary here as a building block upon which to build the logical argument. I'm going to tell you a little about me, so you'll know why I use that word, and how this theory came into being.

8 Mrs. Crook, My Second Grade Teacher

When I was young, in 1966, I had a friend in second grade, Lonnie Holitick, and one day I saw him writing in his math book when the class wasn't doing math, and I asked him what he was doing and he said he was doing math for fun, so I thought oh, ok, let's see if that is true, so I started doing it too, and we finished the whole math book in a couple of weeks on our own, and for the remainder of the year, we did whatever we wanted while our class members did math. That class, taught by Mrs. Crook, was split across two grades; it had second graders and third graders in it, together. That was the first time I had a hint that math was easy. Mrs. Crook was such a delight; she would bring apples every day and slice them so that every student could have a slice to eat, while she read a chapter or so from a book, after lunch, which eventually became enthralling, as the story unfolded.

9 Mr. Jordan, My Geometry Teacher

In the tenth grade, I had arrived with little fanfare, academically, I think to that point, the only book I had ever read was Fun with Dick and Jane, in grade one, only to prove that I could read. The only significant thing that happened in a library was somewhere around the fourth grade, when a classmate challenged me in some way, I responded opening my mouth to smart off, and not back down, and he then challenged me to a fist fight on the playground after school, where he decked me with a single punch and we went to the ground, and came up fast friends from then on. I wasn't the brightest child.

My only other memory in a library was hiding on the back isle so I didn't have to check out a book. As you can see, I hadn't applied myself towards school work in any particular way to this point, in the tenth grade. So there I was, in geometry, where we learned about proofs, and for reasons unknown to me, I solved them easily.

My teacher recognized this in me and gave me extra work to solve harder and harder proofs and I continued to find them to be easy to solve and the class recognized me for it and I found myself standing out and I was a bit taken back. Geometry became my favorite topic to that point in school because my teacher was so interesting, and he recognize something in me that I had not known about myself and he took special interest in showing me who I was, with the sweetest disposition, I simply love you like family. Thank you, Charles Jordan, for setting me on my path; you changed the world, forever; first you changed my world, and now the whole world; like you did for so many, one child at a time.

For the first time in my life, I learned that I had an affinity for something, and it sent me in the direction of science and math. From that experience, where proofs were solved by adding corollary to corollary to build a logical argument to prove a more complex conclusion, that process became a life-long tool in my toolbox. And now, after a few tears, for that dear man who passed away in 2019, Charles Jordan, who taught school for 30 years in Tyler, Texas, but was originally from Hope, Arkansas, born 1936, we're back. I regret not sharing this with him.

10 Einstein's World

I'm going to summarize what Albert Einstein contributed to the world with his definition of a space-time continuum. It goes like this: space and time are connected together; the physical universe that we live in had a beginning, we call that the Big Bang and it will have an end; time itself will cease to exist. He contributed many more fruitful discoveries, but I'm going to add to these.

11 Define Dimension in the Space-Time Continuum

I'm going to explain the definition of a dimension being unique perspective, and see the move from dimension to dimension as perspective gain, because this is the key to all dimensional structure. We begin with the zeroth dimension; mathematically it is represented as a point, it occupies no space, contains no volume, and has no mass; it is nothing. Imagine yourself as that point; everywhere you look around you in any direction you see nothing; you don't even know that you are a point because you cannot see yourself; the whole universe is empty and you wonder how you are looking out, at all.

In order to gain perspective on that point in the zeroth dimension, we must observe it from a point outside that point, and two points define a line. Mathematically, exactly one straight line can pass through any two points. A line is the definition of the first dimension. Imagine yourself as any point on that line, when you look in the direction of the line, all you see is a point, in either direction, one on either side of you, you see a point. Now, you can look at that point and understand that you might be a point, too. You have gained perspective from the zeroth dimension by coming to the first dimension, and you can now say things about the point because you have the perspective gained from moving to the next dimension, but you do not know that you are a line, you only know that you have a point on either side of you, and everywhere else you look all around you is nothing. If you look down the line in this direction, you only see a point; if you look down the line in the opposite direction, you only see a point. The first dimension can be represented in space as length; your line has length, but a point on that line doesn't know that. It can never know that on its own, it must be told by a point that has a broader perspective.

In order to gain perspective on that line, we must observe that line from a point that is not on that line; a line and a point define exactly one plane. Mathematically, exactly one flat plane can pass through one straight line and one point not in that line, that is the definition of a plane. If you are the point not on that line, and you look at that line, you see that it is a line of infinite length, pointing away from you in two directions. You can tell every point on that line that they are part of a line because you have gained perspective on that line. A plane is the definition of the second dimension. You have gained perspective on the first dimension by moving into the second dimension. A plane can also be defined as two perpendicular straight lines. Imagine yourself as the point in the intersection of those two lines. If you look in any of the four directions down a line, all you see is a point. If you look anywhere else, you see nothing. Imagine yourself as any other point on either of those lines, and look at the line that you are not on, and you see the other line, but on your line, you only see the two points next to you. Since you see the line that we named length, you are on a line that we shall name width. Length and width and two dimensions of area, like a sheet of paper; if you are any point on that sheet of paper, and you look around, you see lines everywhere, but you do not know that you are a sheet of paper. You must be told that you are a sheet of paper by a point that has a broader perspective.

In order to gain perspective on that plane, the sheet of paper, you must observe that plane from a point not on that plane. A point and a plane define three-dimensional space, with length and width and now height. The names of the dimensions are interchangeable, and the specific orientation of the three axes relative to your world matters not. The overriding definition for physical space, the universe in which we live, is three perpendicular axes; none of those three lines can ever cross in more than one single point. If I had my 10th grade geometry papers, I could show you that proof using mathematical corollaries. There's nothing terribly complex about it.

Einstein proved that space is connected to time, but if we are going to use the same definition of dimensions, which I have defined here, to prove it, then we must observe perspective gain from three-dimensional space to time. In order to observe three-dimensional space to gain perspective, we must observe change that occurs in that three dimensional space, over time. The same definition of perspective gain is the key to moving along the whole of the dimensional continuum. Every physical object lives in that three-dimensional space, and in order to gain dimensional perspective on those objects, one must observe change over a period of time. Without time, you can look around the universe and have no understanding of what you're observing. Without the context of time, the universe is nothing more than a motionless three-dimensional image; you can't even actually look around. Time is the fourth dimension, which is connected to the three physical dimensions in a linear continuum.

Is there anyone on the planet, who read this book to this point, who does not agree with any of those statements? If you consider yourself a learned person and you're not with me to here, you might as well close this book now, for you're not likely to understand anything else here. I'm not trying to be mean, I simply find myself unequipped to teach you anything. I gave it to you to the degree that I have command of it. If you are young, or find some of it to be less than clear, then please continue to read. Hopefully you will have your own eureka-moment, just like me.

12 Challenge to the Scientific Community

You cannot travel far enough in the physical three-dimensional space to leave the unique perspective that defines it. If you cannot define a dimension better than I just did, then there is no room for any physical space to exist outside of the one I just defined. If you cannot tell a better story than the one I just told, then your theories of multiverse are impossible to be used to define reality and are useless in any theory of everything that is logically sound.

You don't even need the remainder of my theory to come to that conclusion. For decades, I have been amazed at your logical dishonesty, and the gullibility of the whole scientific community who didn't call you on it. You have had these facts all along, all of you could have told that story; for the most part, all of you knew those words; I didn't make them up, you taught me.

Do you even have a definition of dimension as being a building block of reality? If so, then what is it? The entire concept that anyone shall ever travel to another physical dimension is ludicrous because if the definition of unique is that there cannot be another, then there can be no definition of another physical dimension to travel to, by any means, at any time.

Even if there were, there would have to be some parallel three-dimensional space because that is what you are trying to go there with, a three-dimensional physical body, and those other dimensions must be connected to time if it is to be a continuum. So, at the very most, your continuum is three physical dimensions, time and then three physical dimensions, which makes absolutely no sense. You could have infinite dimensions connected to time, in a big wad, but that doesn't provide for a single linear continuum, and therefore it's a useless model. You should apply this model to your own theory and see for yourself where your model fails, and why. No one knows your model more than you and no one cares more than you.

If the dimensional continuum that Einstein gave us did not conform to the model of a single linear mathematical model in a continuum, then we could not use it to predict physics in any way. If his model was a wad, we could not have all of the equations that we currently enjoy in the field of physics. If the definition of a dimension was anything other than unique perspective, there would be no physical science, in the world, at all.

You must throw out Einstein, or you must throw out the multiverse; nothing in the multiverse theory describes reality or uses honest logic. The whole language that is commonly used about such and such from another physical dimension should die here and now; fantasize all you want, but you cannot use the word dimension and call it reality.

I know that my model works because it yields results in every direction I look. What results does yours yield? The multiverse topic doesn't even interest me because I have already dispensed with it; my effort is primarily spent in directions that produce fruit.

Maybe I can find papers that I have written on the multiverse before, and publish later, and if you asked me about it next month, I'd likely come at it from a completely different angle, same result. My world is filled with 10 TB hard drives, and it can be difficult to locate some things. I felt the urgency to release this book and had no time to search for documents.

The bulk of the content in this book is coming off the top of my head in a period of couple of weeks, so it isn't my best work. I was in travail, and this book was born. This book is likely to become the first in a series of books because there is no end to how this model can be applied to gain perspective on reality.

Oh my, I just had more tears, this time with my wife, over a completely different topic. What a season this is in my life. Did I mention that I'm struggling to recover from Covid? I can't recall what I've said to whom. Ugh, travail is exhausting. Let's continue, shall we?

13 Reality is Bigger than the Universe

The physical universe can only contain physical things; that is its definition; physical, three-dimensional, objects are the currency in the economy of the physical dimensions; the laws of physics apply to physical objects in the physical dimensions, and are extended to interact, in a cohesive way that can be predicted, with the fourth dimension, which is time. We have a wealth of scientifically measured results that apply to this continuum to prove it is true; it is not just a theory, it passes the scientific method to qualify as scientific fact.

Even so, that is not the end of the story. You can look inside of yourself to find evidence that you are more than your body; there is more to you than the body that walks around in time. You do not need instruction from anyone to prove that statement to yourself.

If that statement is true, then what else is there that defines you? If there is more to you, than your physical body, and you live in reality, then what defines reality to include the parts of you that you already know exists, even if you cannot name them? If a description of reality does not describe everything that can be named, then it is incomplete; the theory of everything must describe everything, and if it is true, it must do so in a manner that produces fruit.

14 How Many Dimensions do you Function In?

There are more things to life than only physical things; your body, living in this three-dimensional space, at a particular time is not all there is to you; you are more than a body; your mind is not your body; your brain is not your mind, your brain is an organ in your body that is used as an interface between your body and your mind, to make you operate as a seamless whole; your parts are working in harmony within your whole self; you live in there, and you are not only made up of physical parts, you also have non-physical parts; collectively, your seen parts and your unseen parts completely define who you are. If everything fits in a dimension, and you operate in more dimensions than the physical dimensions, then how many dimensions make a whole person?

15 Your Unseen Parts

Your non-physical parts cannot be defined in the physical dimensions. Before we see how they fit into a dimensional model, let's identify the unseen parts of you. We already mentioned that you have a mind, which allows you to think and be sentient. You are not a robot, you make decisions that alter your path, and we call this faculty the mind.

What else can we name? Are you an emotional being? Where do your emotions live? Does anyone here have a heart? We have common vernacular in our language to describe this faculty of a person as having emotions; it is the heart. Why do we call it the heart?

Your physical heart, inside your physical body, pumps blood to your body to sustain its physical life. Likewise, your unseen heart is a pump that delivers the history that you saved there, and that history carries with it a property of emotional content, which varies in degree of emotional weight.

Your physical pump delivers blood, through your veins to other parts of your body, which require it. If there is a parallel with your unseen heart, then to where does it deliver your emotionally-weighted history? It is delivered to your mind.

Your mind has the job of holding your conscious thought for the purpose of making decisions. Your conscious mind is separate from the emotional influence from your heart, but they work as a team to give you the ability to operate in the physical world. Your mind and your heart can be defined by a single word; together, they are your soul.

If you only had a mind, then you would have to solve every problem from the beginning, and it would leave you incapacitated. You would have to examine millions of concepts, from scratch, just to walk across the room. You can look at how difficult it has been to program a robot to do anything, at all, even though the robot has the advantage of human perspective to shrink down their intelligence to a series of commands to function within a small set of confined parameters, and see just how important your heart is to your life.

Your heart holds all of your beliefs; it is your model of the world, as you know it. It is unique to you; it defines your character, and your personality; it is who you are. When you encounter stimulus in your life, your senses take in information, you see something, or you hear something, or you taste something, or you touch something, or you smell something, and that stimulus captures your attention, it becomes conscious thought in your mind, your heart automatically goes into action, you can call it intuition, you can call it an emotional response, you can call it subconscious thought, it is your most immediate source of conscience, whether right or wrong; your heart delivers whatever is has, which you believe, everything that your history has the capacity to have belief about, on that topic or any topic that might be related to that topic, as needed, and delivers it to your mind, so that you can decide how to respond to that stimulus, right now in a complex way, so that you do not have to solve the same situation over and over as you go through life. This is the capacity that allows you to have a complex life on earth, and it is simply miraculous.

You store your beliefs in your heart. Some things carry so little emotional weight that, as you encounter them in the future, your response is seemingly indifferent, though you recognize that you remember or identify with the concept. Other things carry great emotional weight, and you know that they are important, perhaps because they are unresolved in you, or perhaps they mean a great deal to you. They collectively define who you are, and that's why god looks upon the heart.

Your heart and your mind have different purposes, and you need them both. Your soul and your body are collectively called a living soul. Just like your thumb and your palm and your fingers are collectively called your hand, so does your mind and your heart comprise your soul. Your soul and your body have the capacity to make decisions and move about on this earth, determining your own path, and determining the trajectory for your own life. The animal kingdom falls into this category, they are living souls, and they have emotions, and they make decisions, and they have to capacity to go about their business, independently of each other.

You are more than a living soul. You have another unseen part of you that gives you the capacity to receive into your mind, communications from another source, which is external to you, and we call it spirit. Those communications can be called inspiration and holy conscience, as the holy spirit bears witness to your spirit. You hear them in your conscious mind, exactly like you hear your emotional content pumped from your heart into your conscious mind.

You only have one place to deal with conscious thought, and that is the mind. Even the stimulus from physical interaction is sent to your mind. Every communication from every source in your conscious thought occurs in your mind, and that's why we call it the battlefield of the mind. Your struggle to make sense of your world in your mind, that is where you must make decisions about what you believe, and record those beliefs in your heart, so they can serve you in the future.

The very process of growing up, from infant to adult, is tied to the development of your belief system in your heart. It is a gradual process, and it is achieved by making decisions about what you believe, and recording those beliefs in your heart. You are a believing machine, you were created to believe; you are a believer.

The question then becomes, what do you believe? You are responsible for what you believe because they are the choices that you made in your mind, for yourself. They are yours; they belong to you.

They might, or might not align with anyone else, because they are your choice. Free will is your opportunity to make choices for yourself. If you have been given free will, then you and you alone, are responsible for what you believe. Not your parents, not your teachers, not your friends; you and you alone have allowed your beliefs to live in your heart.

Free will is not an unseen part of you; it is a property of being. Your creator has created you, not as a robot, but as a person with the ability to freely make decisions for your own life. You are not a slave; you are a free moral agent. The beautiful thing about free moral agency is that I get to decide where I participate, and so do you. God is not interested in forcing a robot to live with him.

Is there any among you who do not have those unseen parts? Strangely, the scientific world does have cases where catastrophic damage to a person's brain has left that previously fully functioning person with a reduced capacity that exposes this separation of functionality between the heart and the mind, to a significant degree, and it is fascinating to observe. It implies that your mind and your heart each interact through the organ of your brain. Neither of them is your brain, but the dimensional connection between your seen parts and your unseen parts occur in the brain; the rubber meets the road in the brain. You are a seamless whole that cannot function without having a capacity in more than the four dimensions that Einstein defined so eloquently for us.

Who defined the zeroth dimension for us? Was it Einstein? I don't think so; it might be Euclid, who gave us Euclidean Geometry. You math people likely know the answer to that. Really, the unfolding of scientific and mathematical understanding has been a process. Many people have contributed precept upon precept to bring us from the dark ages to where we are today, and I give thanks to them all, but in the end, all truth is god's truth. We're slowly catching on, as more and more truth is being revealed.

16 The Dimensions Beyond Time

We named the unseen parts of you, and now we want to see how they integrate into the single linear continuum of dimensions in our model, this new theory of everything. We saw how perspective gain was used to move from dimension to dimension, from nothingness up through time. Time was used to gain perspective on changes in the physical universe.

Likewise, we shall continue to define the dimensional continuum using the principle of perspective gain. To gain perspective on your changes that occurred in the physical universe over time, we must have a collection of the times for your life, or more accurately, it is the collection that defines you, your heart. To gain perspective one your belief system in your heart, we must use the gatekeeper of your heart, which is you mind, where you make decisions that affect the collection of your beliefs in your heart. To gain perspective on your mind, we must turn to the capacity that transcends your conscious thought, which is your spirit, the source of all things beyond your living soul.

The fifth dimension is the Heart. The sixth dimension is the Mind. The seventh dimension is the Spirit. The unseen parts of you, which complete you as a whole person to operate on this planet, are those three unseen dimensions, which are a continuum added to the three physical dimensions, plus time, to collectively define you in seven dimensions. You are a spirit, you have a soul, and you currently live in a body in this temporal world.

Everything that you are can be explained in seven dimensions. Everything that defines you, which can be named, has a place in those seven dimensions. In them you find all of your thoughts and dreams and everything in the universe. Everything it is to be human is found in those seven dimensions. You are complete in seven dimensions. This is your image. So now what?

To continue to look at the model from your perspective, to gain perspective on your spirit; we must observe changes for all eternity. Your unseen parts are not limited by time. Your spirit and soul shall live after your body dies, they are eternal now. The entire physical universe can pass away and your immortal soul shall live on through all of eternity as the character of you, a spirit being. The only question that remains is where they shall live. The eighth dimension is Eternity. Here is the symmetrical alignment of the seen dimensions next to the unseen dimensions, which you currently have access to:

Time	Eternity
Length	Spirit
Width	Mind
Height	Heart

At the same time, you live in the temporal world, and you live in the eternal world. You primarily live in time, but some day, that part of you will no longer be connected to the unseen parts of you, like it says in James 2.26, "For as the body without the spirit is dead, so faith without works is dead also." We call that the first death. At that time, you will also trade the length, width, and height of the seen world for the length, width, and height of the unseen world. You will continue to function in seven dimensions that complete a person, but now you are in the same seven with god; you no longer have access to the physical dimensions, you cannot be reincarnated, and you are not allowed to talk to anyone in those dimensions, as we see in Ecclesiastes 12:7, "Then the dust will return to the earth as it was, And the spirit will return to God who gave it." After you die, you shall do nothing that is not ordained by god.

Before the universe existed, god lived in heaven. God used the same unseen dimensions that you use; he is a spirit, with a soul, and he lived somewhere. Since god has always existed, the unseen world does not have time, it has a parallel concept, eternity. Heaven is the unseen equivalent of the seen universe in which we live. In the unseen world, there is length, and width, and height, where god lives in his glorified body.

God lived complete in seven dimensions, and when he created you, he created you to live in his image, complete in seven dimensions. He also provided a path for you to be justified and sanctified to prepare you to return to heaven, to live with him in his seven dimensions.

Collectively, the seven dimensions in which god lived before the physical universe existed, plus the three physical dimensions, plus time, comprise a single linearly-connected continuum of eleven dimensions that define a reality in which everything that you can name belongs. This continuum is a model that can be used to describe and gain perspective on everything that has any meaning for your life. In a moment,

I'm going to use it to explain the Big Bang better than it has ever been described.

17 In His Image

This is what god was referring to, when he said that he made us in his image: His complete self, his image, is comprised of seven dimensions; god in heaven, for all eternity, was a complete being, defined by these seven dimensions:

unseen length	unseen width	unseen height	eternity
spirit	mind	heart	

By that same continuum of dimensions, he made us function as a complete whole being in seven dimensions:

seen length	seen width	seen height	time
heart	mind	spirit	

He functioned in the unseen world using his seven dimensions, and we live in this temporal world using a parallel, not the same, but in the image of the seven dimensions like his.

Notice that each individual dimension does not have a specific name, but groups of them collectively form significant parts; when you talk about length and width and height forming our three dimensional space, the names of them are interchangeable; we do not actually know which one is length, but they rightly fit together as a group. So too, does the unseen parts of you, heart and mind and spirit collectively make up the unseen part of you. Again, just as we did in the physical dimensions, where we gained unique perspective as we moved from dimension to dimension up the continuum; so too, do we gain unique perspective as we move up the continuum of our unseen parts.

18 Paul's Prayer Hits the High Points

In this one prayer, Paul talks about the unseen and the seen worlds, heaven and earth; the unseen parts of a man includes his spirit, his Spirit in the inner man; beliefs are found in your heart, that Christ may dwell in your hearts by faith; you can understand the four dimensions of the physical universe, able to comprehend with all saints what is the breadth, and length, and depth, and height; love cannot be comprehended with only your mind, to know the love of Christ, which passeth knowledge. He goes on to say that the process of sanctification is god working in us, and god receives glory for that, which is exactly the . This content in this prayer is astounding.

Ephesians 3:14

For this cause I bow my knees unto the Father of our Lord Jesus Christ, Of whom the whole family in heaven and earth is named, That he would grant you, according to the riches of his glory, to be strengthened with might by his Spirit in the inner man; That Christ may dwell in your hearts by faith; that ye, being rooted and grounded in love, May be able to comprehend with all saints what is the breadth, and length, and depth, and height; And to know the love of Christ, which passeth knowledge, that ye might be filled with all the fulness of God. Now unto him that is able to do exceeding abundantly above all that we ask or think, according to the power that worketh in us, Unto him be glory in the church by Christ Jesus throughout all ages, world without end. Amen.

19 The Big Bang Birthed the Physical Universe

What banged? Who did the banging? From where did it bang? Why did it bang? What was the purpose of the bang? What is the conclusion of the bang? Are there more bangs? Has there ever been another bang? As our understanding of the universe grows, why do the physical laws seem not to hold up when we look at the matter in the universe? Why do we have to invent ridiculous concepts, which have no theoretical basis whatsoever, like dark matter and dark energy, to explain the physical universe? These are questions that science cannot answer, but they are easily described in a way that allows us to gain perspective by using this dimensional model; together with the model, those details and more are available in his word. Let me tell you a story about a bang.

From god's perspective, the eternal god, who has always existed, sat on the throne with his glorified body, in heaven, spoke the words of creation that released a dimensional expansion from the unseen world, to birth the seen world within the unseen world, converting glorified material into physical material, where the unseen dimensions and the seen dimensions coexist in reality, connected in their proper dimensional order. From our perspective, out of nothingness, from the zeroth dimension, at a single point, not an explosion, but a dimensional expansion came forth that contained every physical thing in this universe, with all of the matter that god intended to use for his plan, including the beginning of time, itself, to set in motion the 13 billion year process to achieve his purpose. The length, width, and height of the unseen world are fully connected to and interact with the length, width, and height of the seen world according to the laws that god ordained, such that the dark matter and dark energy are the interaction between them, and their source is in heaven, not our universe. Everything is held together by god's word because he created it; he is the creator who provided for the laws of physics in our world, and the laws of non-physics in his world; heaven can have direct interaction on our universe because it is a connected whole. In a literal six days of creation, in heaven, by his words, he released the process to bring the physical universe into existence, according to his faith. God was probably resting on the Sabbath, kicked back watching the show while the banging was going on, and he has been busy doing all manner of things, besides creating the universe, for the all of the 13 billion years of creation since he released that process to begin. It was all set in motion at the moment the bang began, and it carried with it everything required to accomplish it. There was no need for god to come down here and add the sun in the middle of that process, nor any other thing, because his faith had already accomplished the whole of creation during the literal six days, while he architected the universe in heaven.

God told us that he did the banging, that he banged all of creation into existence, the bang converted part of the unseen world into the seen world, he banged it from heaven, he banged it because he had a plan, and that plan serves his purpose. He said that the universe shall one day pass away and it will be replaced with a new heaven and a new earth, no additional random bangs are going to happen in any fictitious multiverses, and that is the whole of the story because he said so and I believe him. That's my job, I'm a believer. Furthermore, there are no aliens out there in that incredibly vast universe because he already told us how he populated all of the dimensions. Again, he said so, and I believe him. I have not found him in error in any circumstance. He told us the end from the beginning and he has not been proven incorrect in a single thing. Only an omnipotent, omniscient, and omnipresent god can tell the future. He said his word shall not return void, and there is not one shred of evidence that a single word has failed to accomplish his purpose. That is why you can believe that the Bible is the word of god; no man can come close to the complex unfolding of thousands of years, much less billions of years. Our understanding of the meaning of the bible has continually increased over time, the way he intended, and everything I understand is because his plan allowed for his knowledge to be revealed while he is accomplishing his purpose. No man can do that, and there is no other book on the planet that remotely alludes to anything verifiable for future events, and yet the bible does it by the thousands, in such a complex way that we frequently do not fully understand it until we see it come to pass, or he has need for us to see it shortly before it comes to pass. He deliberately does that to thwart the devil's attempts to undermine his purpose. There is a war going on and we have evidence of activity on both sides. I will not spend any time in this book on that topic, but I have published some on the internet, that you can go read for free at foreshown.com

Here's the deal, pickle. You cannot understand all of reality, nor interpret everything within the scope of the universe because there is more to reality than the universe. You cannot force your earth days for creation on god who was not on earth when he did the creating. If you do not have a model that represents reality accurately, you have no opportunity to understand the context of an infinite source. When you understand that both, the seen and unseen, worlds are a seamless whole, with interaction, we can easily remove the need for dark energy and dark matter from the physical universe.

20 The Lamb's Book of Life

From the Book of Life, which existed before creation, for all eternity, and it might gain perspective on eternity itself. We find an eternal god who chose to create a world to share his love. From heaven, his eternal home, he used words, which are the activators of the non-Physical laws that change things by faith, to create a seen world, which from our perspective came from nothing, from his perspective, came from the unseen world in a dimensional expansion.

We know that the Book of life existed before the universe:

Revelation 17:8

The beast that thou sawest was, and is not; and shall ascend out of the bottomless pit, and go into perdition: and they that dwell on the earth shall wonder, whose names were not written in the book of life from the foundation of the world, when they behold the beast that was, and is not, and yet is.

God opens the Book of Life at the Great White Thrown Judgment:

Revelation 20:11

And I saw a great white throne, and him that sat on it, from whose face the earth and the heaven fled away; and there was found no place for them. And I saw the dead, small and great, stand before God; and the books were opened: and another book was opened, which is *the book* of life: and the dead were judged out of those things which were written in the books, according to their works. And the sea gave up the dead which were in it; and death and hell delivered up the dead which were in them: and they were judged every man according to their works. And death and hell were cast into the lake of fire. This is the second death. And whosoever was not found written in the book of life was cast into the lake of fire.

21 THE TWELFTH DIMENSION

The seven dimensions where god lived before the creation, plus the four dimensions he created in the physical universe is the only eleven dimensions that are required for you to understand your place in everything, and perhaps everything you can name could belong in those eleven, but here's why I believe there is a 12th.

Just because god said he has always existed does not preclude him from having a dimension that gains perspective on him where he lives. Forever, is eternity, and there can still be a dimension beyond eternity. I've considered calling it the dimension of God, from whence he came. But here's the thought on that topic that most captivates me. Buzz Lightyear would tell you the twelfth dimension is infinity (and beyond). Eternity already extends from infinity past to infinity future; heaven, itself extend infinitely in all directions, and there is no additional need to express infinity. As attractive as the word infinity is, I have another thought to explore beyond infinity: the twelfth dimension might be the Book of Life.

Functionally, the book is a collection of judgments, god's beliefs naming which people believed, exactly as the heart is the collection of your beliefs, but it seems to be positioned within the three parallel continuums of four dimensions each to be an matched to time and eternity. A proper fit would cause the parallel structure to be consistent, and make the model fit conceptually. I'm not convinced that this accomplishes that, though I will provide an argument for it later.

I've seen perfection in his model that seems to beg for a 12th dimension, which I cannot definitively define, and I'll show it to you in an example below. The most compelling reason that I believe in the possibility of a twelfth dimension is that the ruling structure of the house of Israel has 12 tribes, and a few additional details involving the number twelve. God loves to show us the patterns, which describe his world in terms that we can recognize in our world, throughout the scriptures. Twelve just feels right.

Still, I do have arguments for it being less than compelling that the Book of Life is the answer, including that there is no technical reason, which I've discovered, that precludes God from opening the Book of Life in Heaven, without the need for another dimension; The Book of Life can exist in Heaven, with other books.

Maybe you need to get the courage to ask god about it. I have not asked; I'm still coming to grips with his answer to my last request. He seems to delight in stretching me in ways that I do not know. One thing I know; if you ask, he is entirely capable of answering, and you better prepare yourself for the responsibility of receiving the answer because it can be overwhelming.

Unless you're planning to supplant god as the most high, the twelfth dimension has no bearing on your life. You are not related to it in any way. The eleven dimensions, which I described above, comprise a complete model that is capable of assisting you in the understanding of every useful thing. That is why I can wait until I live with him to hear the rest of that story.

Even so, I will show you why it's alluring for there to be twelve dimensions; that the Book of Life is the twelfth dimension, and these are the twelve dimensions of reality, as my model defines them.

22 Dimensional Continuum from our Perspective

seen length	seen width	seen height	time
heart	mind	spirit	eternity
unseen length	unseen width	unseen height	book of life

23 Dimensional Continuum from his Perspective

book of life	unseen length	unseen width	unseen height
eternity	spirit	mind	heart
time	seen length	seen width	seen height

24 THE LAST DIMENSION

I can see a precious stone in each of those twelve positions. We also know that the Book of Life was used to determine who shall die the second death. Did god have the Book of Life in heaven with him? Or did he use his special capacity to open another book in another dimension? I do not know for sure. If a twelfth dimension exists, then there were eight dimensions that defined all of reality before god called our physical universe into being to make twelve total dimensions.

One could say, and it is my belief, that the Book of Life gains perspective on the whole of the unseen world by containing the judgments over it; it's a direct connection, evidence that god knew the end from the beginning. Part of the difficulty in definitively naming the Book of Life as the twelfth dimension is that we don't know much about the Lamb's Book of Life, except we see it in scripture in the discussion about the New Jerusalem, which you can see later in this book. It describes the creation of new physical dimensions, which come down from heaven, just like our physical dimensions did; the city has twelve angels at the twelve gates, twelve foundations bearing the names of the twelve apostles, the twelve tribes of Israel, twelve precious stones, and twelve pearls, and the inhabitants are in the Book of Life. Even though we don't know much about the Book of Life, my gut says it fits, it follows the corollary of perspective gain, so far as I understand it, and until I have revelation to the contrary, my model names the Book of Life as the twelfth, and final, dimension in the continuum of reality, concluding the theory of everything.

25 THE BIG BANG IN THE UNSEEN UNIVERSE

Just for fun, I'm going to tell a story from god's perspective, using my model. From the Book of Life, which contained judgments, knowing the end from the beginning, sprang forth by dimensional expansion the unseen world that existed for all eternity, where god lived, expressing his inner self in spirit, mind, and heart; and with additional dimensional expansion, he called forth the physical universe to mirror his world, and create people in his dimensional image, with a plan to give them free will, which is the ability to choose to live in eternity with him after their mortal lives ended.

And God saw every thing that he had made, and, behold, *it was* very good. And the evening and the morning were the sixth day. Thus the heavens and the earth were finished, and all the host of them. And on the seventh day God ended his work which he had made; and he rested on the seventh day from all his work which he had made. And God blessed the seventh day, and sanctified it: because that in it he had rested from all his work which God created and made.

What is that Book of Life, that propagative source from which everything sprang forth by dimensional expansion? We don't know.

26 WHAT WE DO KNOW

Death and hell are going to be cast into the lake of fire. Where is that? Is heaven, itself divided into territories, one territory in the presence of god and one territory not be in the presence of god? There is no reason that I can see that this is not the case.

Daniel 12:2

And many of them that sleep in the dust of the earth shall awake, some to everlasting life, and some to shame *and* everlasting contempt.

Eternal space is the unseen world, which perfectly provides for a place to hold eternal things forever. Is hell a thing? I believe it to be a place in the unseen world. As far as I know, it perfectly fits the model. I believe that death ends when the physical universe passes away. According to the model, that means that the physical universe shall be removed, creation shall be destroyed, and reality goes back to the original dimensions that existed before creation. Nothing in the eternal dimensions, in the unseen world, dies. Your spirit and soul shall never die; there is only a question as to which territory in the unseen world it shall live; in the presence of god, or removed from god. The part of the unseen universe that is in the presence of god can be called heaven, and he part of the unseen universe that is separated from the presence of god can be called hell, and the bottomless pit.

27 FROM HEAVEN TO EARTH

Since the unseen length, width, and height of the unseen world continues infinitely in all directions, it is everywhere in the unseen universe. The same is true for the physical universe, the seen length, width, and height of the seen world continues infinitely in all directions, it is everywhere in the seen universe that the expansion has reached in 13 billion years; there is an outer edge, which is continually expanding so long as time exists. You can make an argument that both universes are roughly everywhere, and therefore coexist together, at least within the confines of the current physical universe, which was created in the unseen universe. The closest argument you can have for a multiverse is the two universes, seen and unseen, and in order for a person to travel from the seen universe to the unseen universe, that person must physically die and leave the dead body behind because the unseen universe does not contain physical material. There is some interaction between the two universes because citizens from the unseen universe come to the seen universe and interact, even on a physical level. We do not know the details of how that occurs, but we are told in the bible that is does occur, and indeed, Jesus himself came here, took on physical form, and then returned. However, his case was different from some descriptions of visitations by heavenly creatures, where they come here, interact, and were never born from a human into physical form.

28 THE LAKE OF FIRE

Hell is a literal place because that's how the bible describes it. Since the unseen universe, heaven and hell, is everywhere, and the physical universe must be in it, since it is everywhere, and hell is some territory in the unseen universe, then Hell could be in New Jersey for all we know; hell could be anywhere, and therefore hell could literally be in the center of the earth; it's the same with the bottomless pit. Since the unseen world has existed since eternity past, and it will exist for all eternity in the future, and time exists simultaneously with eternity, even though they are different; so too does the seen world coexist with the unseen world, the unseen world is all around us. The two universes coexist, and anyone with a body that is equipped to live in the unseen universe could eternally live at a location in the unseen universe, which happens to be located in the center of the earth, in the seen universe, and not be burned up because the non-physical body cannot be destroyed by the physical universe. Your eternal torment in hell is likely to be more than just being thirsty, or hot, it is likely to be related to your separation from god. Eternal torment can only occur in the unseen world.

Revelation 20:10

And the devil that deceived them was cast into the lake of fire and brimstone, where the beast and the false prophet *are*, and shall be tormented day and night for ever and ever.

29 THE BURNING BUSH

This same phenomena can explain how Moses saw the burning bush that was not consumed; god opened up the unseen world to let Moses look into it, from his position in the physical universe, and the fire was burning in the unseen world, and the bush was living in the seen world, and the two did not physically affect each other in their respective dimensions. There appears to be a broad range of possible interactions between the two worlds for which we do not have the details for, but clearly they coexist. God spoke from heaven and it was physically heard here on earth. There is a connection.

God didn't take a shuttle bus to travel to the seen world, land on a cloud, speak out loud, and then catch the red eye back to heaven; he spoke from heaven and earth heard it; a direct connection because its structure is a single continuum. Peter walked on water, which can be another interaction between the two worlds; faith gave Peter access to the unseen world. So much of the bible is more easily understood when it is viewed by the perspective gain achieved by applying my model. The star of Bethlehem can easily be explained; god can open the unseen world and display a light in the sky for the purpose of providing a guide to the wise men. Crossing the red sea in the exodus is easily explained. The cloud by day and the fire by night are easily explained. To be cast down is literally to be expelled from dimensions. When Jesus ascended to heaven before their eyes, he didn't simply take off like a rocket and travel in the physical universe to arrive at his cloud or planet or star. The ascension of Jesus was literally him travelling from the seen universe to the unseen universe, right before their very eyes, and he said he will return in exactly the same manner because he is returning from the same place. He goes up the dimensional continuum, and he comes down the dimensional continuum, because they are connected. He can come and go as he pleases because they are connected; he can open the unseen world to the seen world for any reason that suits him.

Do you see that this conversation has no end? The whole bible fits into my model, and everything that you can name fits into it in a logical fashion. As I read the bible, I understand reasons for things it states as true. I have literally read the bible from cover to cover while I was examining it, using my model, in order to find exception or confirmation; nothing in the bible broke my model. The bible tells me what, and my model points to why and how. It's not really my model, god gave it to me. I just keep calling it that so you know what I'm talking about when I refer to it. It's easier than calling it **the single linearly-connected continuum of twelve unique dimensions that form a complete model of our reality of everything, which can be traversed from dimension to dimension by observing perspective gain**. That is the mouthful that I simply refer to as my model, the thing I presented to you here in this book.

30 The Return of Jesus

Luks 21:25

And there shall be signs in the sun, and in the moon, and in the stars; and upon the earth distress of nations, with perplexity; the sea and the waves roaring; Men's hearts failing them for fear, and for looking after those things which are coming on the earth: for the powers of heaven shall be shaken. And then shall they see the Son of man coming in a cloud with power and great glory. And when these things begin to come to pass, then look up, and lift up your heads; for your redemption draweth nigh.

Do not waste time searching the astronomical charts for celestial events in hopes to identify an event with a magnitude impressive enough to point to the second coming because god can display an event is the sky that is visible around the whole world in any way he chooses without needing to plan a 13 billion year asteroid path to streak across the sky and magically circumnavigate the earth enough times for everyone to see it. He didn't need to time special stars at Jesus' birth and he doesn't need to time them for his return. Science is not going to explain those, but my model does.

31 New Heaven and New Earth

Look at the language from the bible; it describes the physical universe being completely destroyed, and what must be a new physical universe being created, which comes from heaven. When it uses the words "down from heaven", it literally can be referring to the dimensional relationship described in the model; from the upper dimensions was once again created some lower dimensions, and they seem to contain physical stuff. Whether the unseen dimensions are the upper, and physical are the lower, or the other way around, is simply a matter of which perspective you are describing it from.

Revelation 21:1

Then I saw a new heaven and a new earth, for the first heaven and earth had passed away, and the sea was no more. I saw the holy city, the new Jerusalem, coming down out of heaven from God, prepared as a bride adorned for her husband. And I heard a loud voice from the throne saying:

> "Behold, the dwelling place of God is with man,
>   and He will dwell with them.
> They will be His people,
>   and God Himself will be with them as their God.

> 'He will wipe away every tear from their eyes,'
>> and there will be no more death
> or mourning or crying or pain,
>> for the former things have passed away."

And the One seated on the throne said, "Behold, I make all things new." Then He said, "Write this down, for these words are faithful and true." And He told me, "It is done! I am the Alpha and the Omega, the Beginning and the End. To the thirsty I will give freely from the spring of the water of life. The one who overcomes will inherit all things, and I will be his God, and he will be My son.

But to the cowardly and unbelieving and abominable and murderers and sexually immoral and sorcerers and idolaters and all liars, their place will be in the lake that burns with fire and sulfur. This is the second death."

32 The New Jerusalem

The second half of that chapter describes the newly created city of Jerusalem in terms of physical measurements, and the most interesting thing is that right here, in the same place that God describes the creation of new dimensions, he connects the twelve tribes of Israel and the twelve precious stones, and twelve gates, and twelve pearls; dimensions with the number twelve have a definite connection in scripture. God doesn't haphazardly do anything; everything is by design for his purpose. There is a connection and the chapter ends with the words, The Book of Life. There you see my fascination with the twelfth dimension being The Book of Life, but it matters not whether it is correct, relative to the usefulness of my model in your life, because any dimension that gains perspective on eternity is not accessible to you, only god has access to it. It remains, the first eleven dimensions, which I defined, contain a model that describes everything you can name, possibly save one, The Book of Life. Look at another connection between the two universes, the glory of God illuminates the city; god lives in heaven, and yet his glory physically affects dimensions that come down from heaven.

Revelation 21:9

Then one of the seven angels with the seven bowls full of the seven final plagues came and said to me, "Come, I will show you the bride, the wife of the Lamb. And he carried me away in the Spirit to a mountain great and high, and showed me the holy city of Jerusalem coming down out of heaven from God, shining with the glory of God. Its radiance was like a most precious jewel, like a jasper, as clear as crystal. The city had a great and high wall with twelve gates inscribed with the names of the twelve tribes of Israel, and twelve angels at the gates. There were three gates on the east, three on the north, three on the south, and three on the west. The wall of the city had twelve foundations bearing the names of the twelve apostles of the Lamb.

The angel who spoke with me had a golden measuring rod to measure the city and its gates and walls. The city lies foursquare, with its width the same as its length. And he measured the city with the rod, and all its dimensions were equal—12,000 stadia in length and width and height. And he measured its wall to be 144 cubits, by the human measure the angel was using. The wall was made of jasper, and the city itself of pure gold, as pure as glass. The foundations of the city walls were adorned with every kind of precious stone:

> The first foundation was jasper,
>
> the second sapphire,
>
> the third chalcedony,
>
> the fourth emerald,
>
> the fifth sardonyx,
>
> the sixth carnelian,
>
> the seventh chrysolite,
>
> the eighth beryl,
>
> the ninth topaz,
>
> the tenth chrysoprase,
>
> the eleventh jacinth,
>
> and the twelfth amethyst.

And the twelve gates were twelve pearls, with each gate consisting of a single pearl. The main street of the city was pure gold, as clear as glass. But I saw no temple in the city, because the Lord God Almighty and the Lamb are its temple. And the city has no need of sun or moon to shine on it, because the glory of God illuminates the city, and the Lamb is its lamp. By its light the nations will walk, and into it the kings of the earth will bring their glory. Its gates will never be shut at the end of the day, because there will be no night there. And into the city will be brought the glory and honor of the nations. But nothing unclean will ever enter it, nor anyone who practices an abomination or a lie, but only those whose names are written in the Lamb's Book of Life.

33 Your Immortal Soul

Everything in the universe came into existence and it will pass away. Time itself will cease to exist. The physical universe shall not last forever; forever can only be found in the unseen world. The soul exists outside of this universe; strictly speaking the soul is defined external to the four physical dimensions, and the therefore the last tick on the clock will have no influence on the life of your soul.

1 Corinthians 15:51

Behold, I shew you a mystery; We shall not all sleep, but we shall all be changed, In a moment, in the twinkling of an eye, at the last trump: for the trumpet shall sound, and the dead shall be raised incorruptible, and we shall be changed. For this corruptible must put on incorruption, and this mortal *must* put on immortality.

Matthew 10:28

And fear not them which kill the body, but are not able to kill the soul: but rather fear him which is able to destroy both soul and body in hell.

34 Words are Spirit

Earlier, we looked at the physical dimensions, where physical objects are the currency in the economy of the physical dimensions; the laws of physics apply to physical objects in the physical dimensions. Similarly, when you look at what is going on in the unseen parts of you, we find that your spirit, mind, and heart all use words; words are the currency in the economy of the dimensions that make up your unseen parts; the laws of faith apply to these dimensions. Jesus speaks to his disciples and tells them that words are spirit, and words are life:

John 6:60

Many therefore of his disciples, when they had heard this, said, This is an hard saying; who can hear it? When Jesus knew in himself that his disciples murmured at it, he said unto them, Doth this offend you? What and if ye shall see the Son of man ascend up where he was before? It is the spirit that quickeneth; the flesh profiteth nothing: the words that I speak unto you, they are spirit, and they are life. But there are some of you that believe not.

Let's take a break from instruction so you can get to know a little about me, and how this book came about.

35 My Personal History

I don't know any Johnson family members; my son, James and I are the last in our line of Johnsons. My father's parents were killed in a car crash before I was born. My dad was a navy man, mechanic, had a double degree in math and physics, fairly intelligent, but for whatever reason, he suffered pride for most of his life and he passed in 2019, shortly before his eighty sixth birthday. In his college years, he had two cars at the same time, and it likely impressed my mother, a small town girl, away from home for the first time.

My mother's father also died years before I was born. My grandmother, Mammy, married a second husband, James, who became known to me as Pappy. Pappy became so dear to me that I named my son after him. They lived in Frankston, about thirty miles from where I lived, in Tyler, Texas.

One of my earliest childhood memories is my mother reading a children's bible to me, like maybe age three or four, and I was fascinated by the stories in it. I remember wanting to be in those stories. Looking back, I believe my heart's desire, as a child, set me up to receive this secret, which I later asked for, and did much later receive. I believe that god looked on my sincere heart, that child, and he began working in my life to produce this fruit, nearly six decades later. I do not know where I would be today without this experience of my mother reading the bible to me. I cannot express emphatically enough how important it is for parents to teach their children when they are young.

36 Salvation

When I was five years old, I went to the First Baptist Church of Frankston's Vacation Bible School. I remember talking to a boy before one of the services. He told me about something that I didn't understand, and I said to myself, "Not me". I had no intention of being affected by that meeting, but I listened to the preacher.

At the end of the sermon, during the alter call, I was convicted of my sins and knew that my path in life was with Jesus, so I went to the front and made my confession before boys. I remember thinking that it didn't matter what anyone else in here thought, I had to do it. When VBS ended, and I returned to Frankston, everyone was so happy in the parking lot. I remember thinking, "What's the big deal? Do I look like a dummy?"

In this same time period, I prayed for a sister. I must have been really suggestible at that age. In any event, I got a sister. If you have a sibling, I suspect you understand. I poke fun at her because she is my sister; in reality, she is exceedingly dear to me to this day.

My childhood seemed ideal until about age six. Suddenly, my parents divorced. My perfect little world fell apart and I was crushed in the process. I cried out to God, "Why has this thing happened? I don't understand how the world works. Please help me understand." As a broken-hearted child, I called upon the Lord and He comforted me. The Spirit of adoption came upon me, and God became for me, the Father to the fatherless.

That experience marked everything I've done since. For most of my life, I thought that it was the loss of my father that caused the huge impact. Decades later, now I understand that calling upon the Lord in that experience set the entire course of my life. I didn't see much of my Father after the divorce, until I reconciled with him when I moved to Dallas in 1990.

Pappy became the most important man in my life. He was the most humble man I have ever met, confident in his purpose, a loving family man, and a faithful Baptist. He worked for forty-two years at Southwestern Electric Service Company, where he served the entire community, repairing everything from the power grid to toasters. I used to walk the line, ride around with him at night and replace the street lights, and discuss all kinds of things. He had a truck full of tools, tool sheds, and warehouses full of stuff. He was constantly teaching me things. He was also the church Treasurer and very frugal. Mammy and Pappy were always giving time and money to their church, and they prayed for Stacy and me every day; the thread of intercession was just part of life.

Mother, being single, worked all the time. Stacy and I were unsupervised for much of our childhood. Stacy was sweet, moderate, and normally social. I was adventurous, mischievous, and anti-social; some might say I was an extremist and a freak, but I hid it well. I investigated everything. I was driven by this need to know everything. Good and evil were somewhat relative terms. While I tried not to harm anyone, very little else was beyond the scope possibility. My motives were basically selfish. I wanted to have a life of my own, and I felt as if the world owed me a chance to do it. We didn't have much, and I knew that if I was going to ever have anything, that I was going to have to go out and get it, and I was employed at age 14, in 1974, for a little over a dollar an hour, and I was thrilled to have it.

Obviously, tenth grade geometry was in here somewhere. Everything to this point sort of seems typical, nothing really stands out, but that's going to change. As unusual as some of these upcoming events are, I just can't tell you about others. I thought these should be told, up to a point; I am deliberately leaving out details and I am just going to tell you what I believe I am supposed to tell you. Actually, I just deleted pages from this section.

37 128

In the summer of 1980, I went to Saudi Arabia to work in the oil field. By this time in my life, I was fascinated with the number 128 to the point that I had to tell people it was my favorite number. It was such an issue in my life I had to have an explanation for it, even though I didn't understand my fascination of it. I recognized that it didn't come from within me, and I was actively searching for anything I could find that aligned with it, in hopes that I could solve the puzzle.

When I arrived in Saudi Arabia, I was told that I would be working on rig number 127, which was so close to my number 128 that I asked, is there a rig 128? They said yes, but that's not your rig. I was perplexed and asked are you sure? I didn't stop talking to them about it for weeks, it just seemed wrong, but I eventually had to let it go. After 100 days in the desert, I went to the Amazon jungle, where I spent my 21st birthday.

38 COLLEGE

I came home from the jungle to two semesters at Tyler Junior College, and then at the end of the following summer, at the last minute, on a whim, I moved to Austin to study Petroleum Engineering at UT. I loved all of the math and sciences. I began to wonder why science could explain so much about the four physical dimensions, but couldn't make physical sense out of the dimensions beyond that. We used the upper dimensions in equations, but no one could describe what happens after the fourth dimension, which is time.

Pappy passed away while I was living in Austin; mother called and I rushed back to Tyler, to see him in the hospital. He seemingly held on until I could get there. I was the last family member to spend time with him that evening, and he passed that night. He wasn't even conscious, but I talked to him anyway. It meant so much to me.

Sometime later, literally without any forethought, I walked off of campus, dropped out of UT during my senior year in 1985, and ended my formal education with a big fat incomplete, which turned into a fail; awesome career move, I highly recommend it. With little else going on, I went seeking the answer to dimensions. I can't tell you why I was so fascinated by it, but over a period of a couple of years, I told people that I was going to solve this problem, long before it actually happened; I'm still friends with them to this day; over a decade ago, they asked me whether I had solved this problem and I told them that I did, and that this book would be published before I die; here it is, old friend; your arguments were invaluable to me, as is your friendship, thank you. Seemingly, as a separate issue, I also began to seek the Lord. Neither ruled my life, but neither was far from thought. For the most part, I was useless.

39 January 28, 1986 Space Shuttle Challenger Disaster

I was not warned about the shuttle's impending disaster; I was focused on the date; for seven years, I was obsessed with the date Jan 28, and I could not figure out why. I told you how it started even before I went to Saudi Arabia, but shortly after returning from Saudi, I knew it was a date, and not just a number. I paid attention to every date I encountered; I even found historical events on that date and recorded them. I could not understand my own fascination with the date, but all of the sudden, on Jan 28, 1986, when the space shuttle exploded while I watched it live on TV; I understood immediately that he taught me how to know when he was telling me something important to him, that did not come from me. My seven-year fascination with January 28 evaporated in a split second. I can show you documents where I collected anything that happened on January 28. In an instant, I never had another thought for January 28, except to remember the lesson I learned. My warning of the disaster was not specific; I had no idea that this particular January 28 was going to unfold in any particular way. God did not cause the disaster, but he used it to teach me a lesson; I had to know when I was hearing from him; spiritual discernment was the first lesson.

40 Eureka!

By this time god had my attention; I was watching about 25 hours of Church TV every week, I was searching, and religious TV fascinated me. One morning in 1987, while seeking, in a eureka-moment, I received this revelation on dimensions, which I bring you now, in this book. In an instant, so many things made sense that I was dumbfounded. I could hardly believe what just happened. I was immediately ecstatic, closely followed by thoughts of implications, which in turn brought about many emotions, including fear, and my face was highly flushed; the impact of it was, in a split second, both exhilarating and terrifying. Over the next moments, as I began to understand its ramifications, it humbled me. My experience was similar to Daniel's experience, in response to being shown a vision, where he says, "As for me, Daniel, my thoughts troubled me greatly, and my face turned pale. But I kept the matter to myself."

Not long after that, I met a girl, and I was 28 years old during most of that relationship, mostly during 1988, so that definitely sets a limit on the timing of when I received the revelation of the dimensions. I think I was 27 years old. I can likely identify an accurate timing if I'd look in my notes. Today, in December of 2021, I'm birthing this book; I need it out of me. I will eventually go dig out my notes to confirm some of these dates and corroborate events, and try to recover specific cases of perspective gain and early beliefs.

41 DIMENSIONS

As a broken-hearted child, I cried out to God, and as a young adult I went seeking, and I knew I was going to find, and I kept knocking on that door until He answered. I have no idea why I knew I was going to solve the problem of dimensions; I didn't even know why it meant something to me; I had no idea it was connected to me needing to know why my world fell apart when my family broke up; I had no understanding that everything in my life had brought me to this moment. Nothing could have been further from my mind than that three-year-old boy wanting to be in the bible.

The moment of revelation was both, scary and euphoric. I began to look for validation and confirmation without telling a soul. I wrote notes about everything. Everywhere I turned, I had an insight about something. My belief system was changing in unprecedented ways, at an accelerated rate. Things were changing so rapidly that I couldn't fully process them. I went through a season of wandering, and wondering whether I was sane.

Chronologically, the theory of dimensions was my second significant experience, and its impact was so great, that I questioned myself fiercely, and I did not tell anyone for several decades. Proving the validity of it to myself was one thing, and preparing to face the world was another matter altogether. I am about to share some things, others still feel too fresh and personal, like an open wound.

42 The Star of Bethlehem

I have seen, with my eyes, a manifestation that could easily have explained the Star of Bethlehem, and many other stories in the bible. It was not a star, it did not act like a star, it was not sitting still, and that is about all I am going to say about its description, except that a similar experience could come in any form, like a burning bush. My explanation of the experience is that we peered into the unseen world to observe something that appeared in our view as though we saw it happening in our world. It was not as shocking to me, as it was for the people with me, perhaps because, to some degree, the dimensional revelation had opened up the reality of the unseen world to me. I intuitively recognized that this was not physical, even though it appeared physical; it looked like a real object in motion, right before our eyes. We did not have a mass hallucination in our minds; we saw something with our eyes, together.

I was in a car with two witnesses who were present with me, who saw it. They immediately became hysterical in the same second we all saw it; I heard them erupt in the back seat while it was going on, we were all looking out the front windshield, as it was easily visible directly in front of the car. It was a brief event, and I drove away, heading home, which was less than a mile from there. Within a minute, more like fifteen seconds later, I turned around, while I was driving, looked at them in the back seat, and tried to talk with them, but they were still hysterical and huddled together. I inquired again and they acknowledged to me that they saw it; I asked a direct question and got an "uh huh", and that was the end of it.

That was the only exchange that we ever had about it. After we arrived home, less than two minutes later, we settled in and I tried to bring it up again, but they refused to talk about it. We all spent the whole night together, as a group with other friends, who also came over just hanging out. We have not spoken about it again, to this day, even though we are still friends. It is clear that we all experienced something together, which is not a common experience to most folks. No matter how you explain it, your description is going to be way out there.

After a season of diversion, one night, alone in my room, the Holy Spirit convicted me. I fell on my face and cried out to God. I recommitted myself to Him, and He comforted me. The next day, my girlfriend came to my door, but I did not open it. I only yelled through the door for her to leave. It was a terrible thing to do, but I had to turn from my entire social scene, and I handled it extremely poorly. I faced the reality that I was a mess and God was all that mattered. I began to seek Him directly, not peripherally, by my own choice.

43 Someone Walked Through the Wall

Some weeks later, around 1989, after falling on my face and changing my social world completely, I was lying in bed, when someone walked through a wall right before my very eyes while I was wide-awake and alone in my bedroom. To be more accurate, they walked through that same closed door, which I refused to open for my girlfriend, without opening the door. Someone walked through my bedroom door, literally, without opening it, and sat down in a chair next to me. I could not believe what I just saw, and I did not know what to do.

I did not feel threatened, and I never thought I was going to die during this experience, but, after seeing him walk through that door and sit down in a chair next to my bed, three feet from me, I had to turn my head away, I had to stop looking. I simply couldn't deal with it in any other way. I never opened my mouth and I regret, to this day, that I did not have the courage to say something. At the time, I resigned myself to silence because my thought, in that moment, was that anyone who can walk through my wall is entirely capable of doing the talking, and yet all I received were impressions, nothing audible. After what seemed like ten or fifteen minutes, I began to calm down, turned to look and he was gone. I did not tell anyone for decades.

Perspective gain continued to come, and a major concern was the grandiose statements that continually came to my mind. Was I delusional or insane? I exhibited many of the same traits as some mentally unhealthy people, and in part because I really was really a mess. I spent more than a little effort in that direction.

So many topics had become important to me. I didn't have time to research it all, so I watched every documentary I could find, regardless of the subject. The various documentaries showed me that the revelation allowed me to gain perspective in almost any discipline.

I could see where a field of study was going, without having any prior training. I knew answers before researchers expressed the questions. I watched many fields progress over the years, and confirmed many insights.

I also continued to watch all of the religious TV I could find, one of which was Larry Lea and Church on the Rock, COTR, in Rockwall. Around 1990, I drove to Dallas, from Tyler, twice a week for a year to attend COTR. Then, I reconciled with my father and moved in with him in Dallas. A church member recommended me for a job at the company where he worked.

The COTR years were so important. Spiritual authority and many other things were put in place. I continually read the Bible and recorded insights, the most significant of which was that I read the bible from cover to cover and everything aligned with the model; for the first time, I found a source that directly dealt with it on a concrete level; the bible and the model were meant for each other. I think the bible qualified as the fourth book that I can claim to have read in my life; maybe it was number three, I would have to confirm that, I still have I believe four of the books; Fun with Dick and Jane, from grade 1, has escaped me. I expected God to direct me to submit the revelation to COTR, but He did not, so I did not release it. I was at COTR while Larry Lea was Pastor and remained there for a long time after he left. The whole experience of that transition, him stepping down from being pastor, sent me to a strange place. I became a lone ranger, not submitted to any authority.

Years passed, in which significant spiritual things happened in my life, which I do not minimize the importance of, I simply have decided not the share them for brevity. I am not really trying to tell the complete story of my life; I am just hitting some high points. I want you to get some idea how it unfolded without you focusing on me. I just want you to know that I can tell the story because it is not a screenplay that I made up one summer; I had to grow through this, I am still growing in this.

For several years, including 2002 and 2003, and for an extended season, I was knit together with my sister's church. The following experiences, and so many important things, happened during that time. Some were astounding.

44 Instantaneous Physical Healing

I was playing the drums in church, and my wrist began to hurt, as usual. I had known that I needed carpel tunnel surgery for a long time. After about 15 minutes of playing, the pain always made playing a chore.

On this particular day, when it began to hurt and, without thinking, I just shouted, "heal it Lord", and the pain went away immediately. Later, I repented for speaking to Him like that. It is not my habit to command anything of him, and I really did not understand what came over me. Today, I still play drums without pain.

45 The Rush of a Mighty Wind

In 2002, I was alone in my bedroom, sitting at the desk, working at my computer, where I am sitting right now, as I write this book. All of the sudden, I heard this indescribable sound, aloud with my ears, and began to stare in the direction of the sound, to my right; I began to stare at the closed window blinds next to my desk for a brief moment, for maybe fifteen seconds or more. My mind tried to make sense of it, but could not. It was so intense that I stared at my bedroom window, expecting it to blow in and kill me on the spot. I thought I was going to die, right then and there.

Whatever it was had me right in its sight and, no matter the outcome, there was nothing I could do about it. It was so compelling that I could not turn away. I was frozen, not scared stiff, more like enthralled, staring straight at the closed window blinds. I wondered whether a jumbo jet was about to hit my house, and then whether a nuclear missile was about to destroy the city, and then whether a huge tornado was about to blow my house away, or a train. The sound kept getting louder and stranger while my mind tried to explain it away, and I kept ignoring the still, small voice, which was saying this phrase, "the rush of a mighty wind" and then it stopped. I almost fell over. I was highly excited and the adrenaline was pumping.

When I realized that I did not die, I thought whatever it was must have flown over the house, so I ran out the back door of my house and waited to see the explosion, looking in the sky expecting a mushroom cloud to appear on the horizon, and yet nothing happened. Puzzled, I thought maybe it was not that big and it went toward another part of the city, so came back in the house and called a friend that lives about fifteen miles in that direction and asked her if she heard any big explosions in her area. She seemed to humor me, as I tried not to be alarming while I asked increasingly bizarre questions.

Then, I called my father, who lived three miles from me and asked him whether he heard anything. He said no. He could tell that I was completely out of sorts. I could tell that he wanted to be there for me, but he just did not know what to say. I tried to talk to him long enough to let him know I was all right, so he would not send the straightjacket squad. We laughed about how worked up I was and I tried to calm down, realizing that the experience had passed.

Even then, I could not let it go. I heard children playing in the neighbor's front yard, so I went outside and asked if they heard any big noises. They only mused at my inquiry and politely said they had not. The further this went without any reasonable explanation, the greater it affected me.

I could accept a nuke, but "the rush of a mighty wind" just would not do. I was running out of explanations and it became clear that this experience was not from the seen world. I was the only one who heard it; it was audible, not just in my mind. At that point, I had ruled out all but one rationalization; the rush of a mighty wind was the Holy Spirit. I could not bring myself to believe it, but my mind had run out of alternative possibilities. I marveled at it, and went back to what I was doing before it happened.

46 God Spoke to Me, Aloud, I Heard Him with My Ears

In 2003, I went with around twenty church members on a mission trip to Uganda. I was generally lending my musical talent and recording skills, so mixing the music, which we recorded during church services was an ongoing task for some months. I would take the recordings from the board, in the sanctuary, home and mix them down into a CD. We eventually got everything functional at church, but before that, I performed the mixing in my recording studio at my home.

One day, after mixing a worship CD, I was walking out of my studio, reached for the doorknob to open the door, had my hand on the door knob, facing the door, and from behind me and above me, God spoke out loud to me, audibly to my ears, aloud, and it was so real and frightening that I turned around and answered, "You're crazy!" My big moment had arrived and all I could muster to say back to him was that he was crazy. There I was, talking to someone that I could not see, and it was so impactful that it changed my perception of the configuration of the building, which I constructed with my own hands, forever. In my mind, there is no longer a ceiling above the door, where I turned around expecting to see god sitting in a specific place; that unbelievably stunning voice came from a specific direction.

When I go in the now, and go to that same door, grab the knob, and turn around, I am always surprised to see the ceiling right in front of my face. In an attempt to reconcile what I heard with my ears, my mind had to change what I believed to be the reality of the building. It was more real for God to be sitting in an open area than it is for the ceiling to exist. There is a vaulted ceiling in the room to the right of that spot beside the door, and the experience rather makes me think that the door is actually located in that other room, because you can at least look upward from that location. It is all somewhat confusing to me now, in a building that I built. The two worlds collided and the unseen world prevailed.

47 The Voice of God

There is no way I can accurately describe my impression of God's voice. It was loud and big, but that is not the point I am making. The tone of his voice revealed his character, even if you ignore the content of the message. His majesty, his compassion, his love, his absolute power over all, was completely obvious. Those and a thousand other character traits were all immediately and individually discernable in a split second.

I can assure you that God will get your full attention when He speaks aloud to you; He is used to getting His way. God's voice included a matter-of-fact tone blended with personally engaging; authoritative, yet not imposing; powerful, but not threatening; so many things all at once. Kind and loving, no nonsense allowed, so many textures all together. God's voice concisely conveys that He is the master of the universe with no effort at all. I had previously believed that God had a sense of humor, but there was no trace of it in his tone. He sounded like nothing could surprise Him; your wit is not going to make him giggle.

48 Fear God

Nothing deserves to be feared more than God, himself. I have heard the voice of God, out loud, with my ears; just like I talk to you, with my mouth and you hear me with your ears. It was the most terrifying thing that I have ever considered in any thought for my entire life. Nothing compares; everything pales in comparison.

Nothing could be more humbling than personal interaction with God who created the seen world with his words. If all you expect is a cuddly, loving god, you are in for a rude awakening; you are going to be shaken to the core, every one of you, no matter how righteously you lived. Every knee bowing is an understatement.

Nothing in heaven or on earth can adequately prepare you for the awesome reality of answering to God, face to face, for what you have done with what he gave you. It will be the most frightening experience that you shall ever experience. No imagination in anyone, for the entire history of the world, can name anything more frightening than simply hearing the voice of God out loud, except that voice also be used to condemn you.

49 You Too

Why would God, the creator of all things, speak to me? He is no respecter of persons, yet we do not hear about Him talking aloud to everyone, so what would bring about such a thing? Have you heard of anyone else alive today that has said that god spoke to them aloud? I have not. Oh, there are plenty of people that say god told me this and god told me that, but I am not aware of anyone alive today, besides me, that claim that god spoke to them such that they heard his voice with their ears. Why would anyone claim that? Yet, here I am.

All of those experiences, which I listed, are things I have personally seen and heard with my eyes and ears, not some mental imaginings or spiritual communications. Those are my real world experiences, taken in by my physical senses. I did not physically feel any of them; I did not touch any of them, or smell, or taste anything.

With each experience, time itself very nearly stopped, my heart rate instantly shot through the roof, beating a mile a minute. My mind raced through entire conversations in a split second. everything in me was hyper-alert because I realized that the extraordinary was occurring and had become my reality, exactly like one might experience in a car crash at the moment you realize that you might die and you have no control over it whatsoever. The main difference between these supernatural experiences and the car crash scenario, which I have also experienced, is that in these supernatural experiences I never felt threatened, except for the one about the rush of a mighty wind, so in general, it was not fear for my life that I was feeling, but they had my utmost attention. These were life-flashing-before-your-eyes type of moments without the thoughts of dying, in particular.

This is an incredible story, by most standards. From my perspective, there are only a few assessments that can apply. 1) I could be completely delusional, where I believe a lot of weird stuff that is not true. 2) I could be deceptive, making up a story that I do not believe. Or 3) I could be telling it like it is. The unfortunate part for you is likely to be that no man can prove or disprove most of the content. I know because I have spent decades unsuccessfully trying to prove or disprove it. I believe you can prove it for yourself, but you are simply not likely to find the scientific community, or any other community, beating a path to your door to confirm it.

The case that is the least likely to be correct is number two, me being deceptive, for multiple reasons. To begin with, I am simply not that imaginative, but you do not know me, so I do not blame you for not discounting that one. However, the main reason that I am not deceiving you is because the bible specifically warns against deceiving believers, in particular, and as a Christian, that particular point I take extremely seriously. To violate that on a grand scale would be the equivalent of not believing at all. There is no way to reconcile me being a believing Christian, performing mass-deception of believers as my life's mission, and then think that I'm going to have a warm welcome in heaven. Those thoughts do not even make sense in a deluded mind.

That leaves us with the opposite ends of the spectrum; either I am crazy or I really can recount an amazing story. The supernatural experiences are not the point. They are notable moments along my path, important and meaningful to me because I grew through them, but I contend that their significance to others is minor in comparison to the model presented within. There is a point to all this; I did not just randomly cross paths with creatures from another world.

I will talk all about my supernatural experiences and you can poke holes all you want, be entertained, or whatever it does for you, but the model can change your life. I have been irreversibly and fundamentally changed through my experiences. My perspective has been altered and the genie can never go back in the bottle. My view of the world is significantly broader than it used to be. My belief system was expanded to a more complete picture of how things are related. Everything that you can name has a specific place in relation to everything else, and the model can be used to gain perspective on everything, to put everything in its place. This, in fact, is the theory of everything. It's the holy grail of science. If I can gain perspective on everything, then I should be able to shed light on some amazing stuff, and downright answer many of life's hardest and most fundamental questions, and I can do that; you got a taste of it in this book. I can assure you that I stand in judgment for it all, whether good or bad, true or not, witness or deceiver. You be the judge.

God is no respecter of men. He wants to do things through each of us, but we have to be willing. God tested Abraham, and Abraham said, "Here I am."

God spoke to Israel in a dream, "Jacob, Jacob, and he said, "Here I am."

The Lord saw that Moses turned aside to look. God called to him from the midst of the bush, saying Moses, Moses, and he answered, "Here I am."

The Lord called to Samuel, and he said, "Here I am."

The Lord called to Ananias in a vision, and he said, "Here I am."

He called to me, and I told him he was crazy, but if you look in the eighth paragraph of this book, I eventually said, "Here I am."

He has a purpose for me, just as he does for you. Prepare yourself to say, "Here I am."

50 Such a Time as This

In my life, I have directly experienced multiple interactions with the unseen world, which are not typical experiences in the seen world. You have to decide whether I am a lunatic, or honestly and accurately describing my experience, as there is no other explanation.

Which is more likely? Seeing a man walk through a wall, or hallucinating a man walking through a wall.

Which is more likely? Seeing a chariot of fire, or hallucinating a chariot of fire.

Which is more likely? Audibly hearing the voice of God out loud, or hallucinating the voice of God.

Which is more likely? God giving you a theory that all the world's scientists cannot disprove, or you making up a theory that all the world's scientists cannot disprove.

All of these, and more, are my experiences. If you had asked me those four questions before I experienced them, I would have answered hallucinations to the first three, and then wondered about the criteria for the theory before passing judgment, and if sufficient evidence for the validity of the theory exists, then I would have to say that that sounds more like a God thing than man thing, for what kind of man does that? What kind of man outsmarts the whole world? Einstein? Who thinks they are like Einstein? Nobody, and certainly not me.

Our natural inclination is to put things into perspective according to our daily experiences, and those particular four are not common experiences. If any of these happened to you, you would wonder whether you are delusional. You would question yourself, and analyze yourself. No matter how convinced you are in the moment, of the validity of the experience, you will undoubtedly question yourself at some point after the experience. I know this because I did. I found myself involved in circumstances that my own imagination would not dare to dream up, and had to deal with them. When things like these begin to happen, you feel like a survivor that just had their world destroyed by a disaster. In fact, your experience of reality just ran off in a completely new direction. Everything that you knew before is suddenly affected.

I now know that these experiences trained me for such a time as this, by design. I am exceedingly thankful that I did not abuse the confidence; that I kept this secret for so many decades so that god could complete this work in me. I had every opportunity to use this in a selfish way, but I did not. My mission might, or might not, be complete. That is not for me to say, but I can rest easy now, knowing I have given all of the glory for it, to god, our creator, who created our universe, for he alone is worthy. He looked upon this child and weaved me into his story, and I worship at his feet.

51 Supernatural Experiences

I do not believe you should seek experiences. You should invest yourself in your relationship with him. The closer you get to him, the more equipped you become to participate in his world. It's really that simple. He's a person, and you get close to him exactly the way you get close to anyone.

The word supernatural literally means more natural, and above natural, both of which make sense in the dimensional model. The unseen world existed before the seen world; it was the full definition of reality. Everyone that lived in it naturally understood it. After the seen world was created, naturally, they still understood the unseen world better than the newcomer; the unseen world was more natural to them than the seen world. Likewise, the seen dimensions are lower in the dimensional continuum than the unseen dimensions, which makes them the subclass, where the unseen dimensions are then the super class. Strangely, the unseen world always functioned by their own laws, and the seen dimensions literally had to be the lower dimensions in order for the mathematics to work out, which give us the laws of physics in the natural world; the dimensions didn't change numbers, just because we got more. Basically, you can say that god knew from the beginning that the physical dimensions were going to have a place in reality; it all fits perfectly together, even though in eternity past, there were no physical dimensions.

It's important for me to share some things about my experiences because they aren't common in the world today, and there's no reason that they aren't common, other than unbelief. This account will bring attention to the exercising of faith, and I have no doubt that they'll become increasingly common. We say prayer changes things, and they do, but faith is way bigger than asking god for help; faith is the more natural way to bring about change.

For example, you can pray and ask god to heal you, and god can choose to exercise his faith to heal you, or you can exercise your faith to heal yourself. It's the exercise of the same laws that god uses. He said reality on earth is subject to the way you say it is in heaven; the same laws of faith that god created the universe with are available to you, if only you believe. He's not in heaven being a traffic cop to determine whose faith is allowed to work, any more than he's picking and choosing which apple falls to earth under the influence of gravity; laws are laws, they always work within their context.

The implications of this becoming widely understood is that exercising of faith that affect the seen world will become common place, and it will appear to many that wizardry on earth has suddenly left the pages of fiction to become reality. Miracles will be more visible, they're happening now, they just don't get much exposure. Just like exercising physical muscles increase physical strength, exercising spiritual muscles increases spiritual strength.

The world has had faith healers, for example, and some had significant results. People have received revelation in some area and proved that they could participate in that economy. I've heard stories from Jewish tradition about Jesus in his childhood, killing small birds and bringing them back to life. He didn't just suddenly get zapped by god at his baptism to receive the gift of miracles; he practiced faith and grew in faith over his lifetime.

When you first know that your words have changed your physical world, by faith, in a way that is not explainable with physical laws, you're changed forever. There's no going back; you can't change my mind; you can't put that genie back in the bottle. Especially after you find out that god isn't granting wishes, that faith can be used for good or evil, exactly like your physical actions can do either, you're changed forever. The world is a different place when your moral compass, and perhaps the lack of exercise in faith, is all that stands between you and a world that no one recognizes, you're changed forever. It's the equivalent to suddenly having your finger on the nuclear arsenal; you have nukes at your fingertips, and if you do, then so can others.

There's nothing special about me. I fall from a tree to the earth because the law of gravity works; it works for you just like it works for me. I change things on earth because the law of faith in heaven works; it works for you just like it works for me. It works for me just like it works for god. The only difference between his capability and god's is what he can believe that I can't believe. It's the same law, I haven't spent an eternity exercising mine, and I'm not omniscient, so I'm clearly not on his level, and I'm not going to ever get there, any more than I'm going to exercise my physical body enough to look like a professional body builder. However, if I apply myself, I can progress in either of those directions, to some degree; physical exercise is profitable, and spiritual exercise is profitable.

Everyone isn't going to move forward in faith at the same rate because everyone doesn't have the same revelation. You are able to believe because you have been shown. The difference in abilities will result in a few people taking the lead and becoming notorious for it, especially when they show off and boast; they claim the glory for it, and it will be evil, exactly like murder.

When I recognized this possibility in my own life, I recoiled from it, it scared me, and I chose not to exercise it because I was afraid of me, I was afraid my selfishness would be too tempting, and I didn't want my finger on the button because I was afraid I'd push it. With all of the things I've experienced, one might think it would be easy to step into, but it's not. Ask Peter about walking on water. If you talked to people who were in my life around these times, you'd hear some pretty strange behavior. No one really knew what was happening to me, but some could see peripheral effects.

I have this stick, about six feet long, that a beaver cut and cleaned. I've been carrying it, in my possession since around 1986 or 1987. I still have it; I saw it the other day. It's like a staff, very strong and lightweight, great for hiking. I have stared at that stick believing I could turn it into a snake, like Moses did for Pharaoh, and I think the main reason I didn't was because I didn't want a snake that large running around in my room; I didn't have a reason to exercise my faith in that direction, but if I did, I was ready. I did not try it because I believed I would be successful. I know that sounds fanciful, but the point is, my reality has changed; I no longer look at the world like most folks, and there's nothing you can do to change my mind.

I found a way to live in both worlds. I'm comfortable with me. It's really easy for you to look from the outside and say I should have done more in some direction, or whatever, but you haven't walked in my shoes. I understand you thinking that way, and it doesn't bother me. I even agree on some level, but my walk through creation is between me and my creator. I'm completely at peace with you thinking whatever you think about me. I know to whom I answer.

One day you shall see this reality of which I speak. I might not be here with you, but this reality is described in the bible; it is coming. The fact that I published this book is going to be a factor in accelerating the rate at which the world can believe, and some will advance in faith more quickly because of this revelation in their lives. Some will change the world for the better, and some will change the world for the worse. Increasingly, our grandparents will not recognize this world. I believe that time is near, and I've put my money where my mouth is, much to the chagrin of some around me.

So I don't tell you of my experiences to puff me up. I leave out details so I don't get puffed up in your mind. I do not seek glory. I've hidden this reality, in varying degrees, for decades. I didn't expect to give this to you now; a month ago, to the best of my understanding, I was no closer to delivering it than I was decades ago, even though I knew that someday I would give birth. I didn't see it coming, this was premature in my eyes, but it wasn't up to me, it had to come out now. I have no idea where it will lead, but that isn't my responsibility. I'm doing what I'm supposed to do; what happens with it from here is not my responsibility; I'm fulfilling my mission. You can reject it and I will fade quietly away and be completely satisfied in the knowledge that I carried out my mission. If you want to talk about it, I'm here. That's really all I have for you. I failed like a blubbering fool before, but I am not that person now. If god wants to use me, he knows I will say, **here I am**.

Ok, enough about me, let's use the model.

52 THE SANCTUARY

In the bible, god describes the tabernacle, the temple, the sanctuary, in a very precise manner because it is useful to us. In the model of the sanctuary, using the furniture, god describes the path that Jesus took to leave heaven to meet you at the cross, and the path for you to follow Jesus back to heaven to live in the presence of god. As it happens, because god knows the end from the beginning, the sanctuary building model is also an accurate model for the proportions of your inner man, and it also directly supports my model of dimensions, which I give you in this book, as being the theory of everything.

God dictated the dimensions of three tabernacles; the one Moses pitched in Exodus 36 and 38, then in I Kings 6, and in Ezekiel 40 and 41. There is much to glean from these models, but for now, I only want to bring your attention to the dimensional relationship between the holy place and the most holy place.

The following diagram is from the tent that Moses pitched. It shows the holy of holies as ten cubits by ten cubits, and the holy place as ten cubits by twenty cubits. The relationship between the measurements, given by god, the dimensions of those two spaces, relative to each other, is that the holy place is twice the size of the most holy place. If we look at my model, the spirit is a single dimension, and the soul is comprised of two dimensions; exactly the same dimensional proportions that god told Moses, two to one.

You are a spirit, you have a soul, made up of your mind and heart, and you live in a body. In the diagram, the outer perimeter represents your body, and your inner man is represented by the holy places. God gave us an image of the way he is made, in the sanctuary, and it is also an image of the way we are made in his image. We have the same parts because we are created to function as he functions, and eventually live with him in his world.

Without having any significant understanding of the tabernacle, there was a moment in my life, where I popped up out of bed and went to the bible, knowing that I was going to find this relationship. I didn't happen across it and then commandeer it to support my argument; with no forethought, I went looking with the expectation that I would find confirmation, and found it exactly as I expected. Is it coincidence or do we see meaning in purpose?

                    50

              ┌─────────┐
              │   10   │
              │ 10   │
              │   │
              │   │
100   │ 20   │
              │   │
              │   │
              │   │
              └─────────┘

**The Dimensions of the Tabernacle of
the congregation that Moses pitched**

53 THE TRINITY

I know I've written on this topic in the past, and I hope to someday publish my previous writings, but again, I'm going to give you my current thoughts, off the top of my head, as is everything you have read so far in this book. The trinity is not in the bible, it is a modern human construct. I can't tell you the history of it, but I do have an opinion on it. I don't see any particular damage caused by men attempting to understand the bible, and give it descriptions in an attempt to discuss it more clearly with each other, so long as it doesn't come in conflict with that which is specifically expressed in the bible. In fact, my model does exactly that. What I'm saying is I have no issue with you referring to them collectively by any term you established to agree on, within certain bounds. Likewise, I happen to have my own interpretation and description, according to my model, which follows.

It seems to be more than a coincidence that god expresses himself as three individuals, the father, the son, and the holy spirit, and those three expressions line up with the unseen parts of you, and yet there is only one god. If god is a trinity, then you created in his image are also a trinity. God is a spirit, the holy spirit, and you are a spirit. The heart of the father contains all truth, and your heart contains your truth. The mind of Christ, able to rightly divide the word of truth, is your standard, and from it Jesus is the word of god, and his words created our universe; Jesus represents the conscious part of god, the mind. God has a mind, expressed in Jesus, and you have a mind that makes your declarations of faith and determines what you decide to believe, which is hidden in your heart. The three expressions of a triune god are exactly expressed in you, a triune creation, created in his image. If god is a trinity, then you are a trinity, but god is one god, and you are one person. God expressed himself in three persons, so that we can understand the parts of a person using language that we can relate to, while the parts fit together functionally to accomplish a specific purpose in defining the whole individual.

God could have been more specific in the bible, but he chose to veil the truth for his purpose. Everything related to god is related to his purpose. The central topic that you should be focused on for the whole of your life is his purpose for your life.

God never told us to use the cross as the symbol for him, or the fish for that matter. Like the trinity, that is something we did. I don't find any particular objections to it, so long as it is only a reminder of him, and it is not an image that we worship. Only god deserves to be worshipped. If I find a direct link to the cross being instituted by the church as an image to be worshipped, I will stand on the opposite side, directly opposed to its use. Currently, my research has not been in that direction and I have no revelation of it.

My commitment to god is to change my belief on any topic when new revelation sheds life on truth that is new to me; I go with god's truth when it's revealed to me; I choose to hide his truth in my heart, so that I can believe as he believes, and I can use that truth to make good decisions for my life. If we choose to hide lies in our heart as our belief system, we will make poor choices. The more truth we believe, the better our decisions.

By profession, I've been a computer software developer for the past couple of decades, and we have a saying, garbage in yields garbage out; if you input bad information into a computer, it has no opportunity to produce anything except to give you bad the bad information. There is no capacity within the computer to recognize your incorrectness and give you back a corrected result. Your heart works in an identical way; it only knows what you decide in your mind.

In fact, the reason that we have a computer, at all, is because it is a mechanical device that we made in our image; we have used the only model of intelligence, which we know, to operate like we operate, to the best of our ability, in order to be useful, and that model is us. I could go through the whole parallel, and maybe I will another day; this baby is about to be birthed, and I can't stop it. Much of the remainder of this book is things that I have previously written, some have been touched, but it's likely to be less cohesive to the book than the chapters down to here. I apologize for that. I hope to one day make it right for you. Let's depart from what I have to say for a moment, and hear what Einstein had to say.

54 Quotes from Albert Einstein

"I didn't arrive at my understanding of the fundamental laws of the universe through my rational mind."

"The intellect has little to do on the road to discovery. There comes a leap in consciousness, call it intuition or what you will, the solution comes to you and you don't know how or why."

"We are souls dressed up in sacred biochemical garments and our bodies are the instruments through which our souls play their music."

"When you examine the lives of the most influential people who have ever walked among us, you discover one thread that winds through them all. They have been aligned first with their spiritual nature and only then with their physical selves."

"One thing I have learned in a long life: that all our science, measured against reality, is primitive and childlike. We still do not know one thousandth of one percent of what nature has revealed to us. It is entirely possible that behind the perception of our senses, worlds are hidden of which we are unaware."

"I'm not an atheist. The problem involved is too vast for our limited minds. We are in the position of a little child entering a huge library filled with books in many languages. The child knows someone must have written those books."

"Everything is determined, every beginning and ending, by forces over which we have no control. It is determined for the insect, as well as for the star. Human beings, vegetables, or cosmic dust, we all dance to a mysterious tune, intoned in the distance by an invisible piper."

55 ODE TO STAR TREK

Captain Kirk's mission, to go where no man has gone before, is how I live my life. Not because I'm a Star Trek fan, but because the thought of exploration allows me to dream a reality bigger than me. That's a significant reason why developing software is such a good fit for me, the application doesn't exist anywhere, no one has gone there, and that's where I live.

I didn't pick software development because it paid well, I simply had to have it in my life, and a career just sort of came with it. I steadily, daily, travel roads that have not been travelled. I don't just want to, I have to. I would not feel alive without it. Inventor is not a job description, it's an addiction. Exploration is not an activity; it's a drive that isn't equally distributed throughout the species. Some could more easily stop breathing than stop exploring. I'm one of those. I seldom complete an excursion before being ripped away to start another; my attention is more than captivated, I'm consumed.

I'm completely enthralled, my focus scarcely capable of sustaining life outside of it. No one understands why I am driven, as I have hidden it as best I can, and to come out of the closet is to be limited. I know no limits and I'm not looking to know them. I'm not trying to fix it, it isn't broken. It's who I am. The dark side of it is this; I'm a terrible finisher because I can't stop starting the next thing. I suspect my wife would comment on my poor follow through, regardless of my noble intentions.

56 Life in Your Dreams

My recollection of very early child hood dreams include a recurring one that I couldn't wait to get to sleep in order to experience another episode, which involved small people in my hands. They were so cute and I took care of them. The best I recall, these were the beginning of lucid dreams, where I had some measure of control within the dream, or at least could anticipate another adventure.

Other typical dreams included flying in my house or over the city, and the normal things one hears of, like being naked at school. Most of which I consider to be simply part of everyone's dreams as they grow up, and attach no particular spiritual meaning to, apart from the natural function of dreams.

I have no history of nightmares. As a young child, I eagerly investigated anything that the world proposed as scary material, just to prove that it wasn't really scary, and nothing registered as being particularly upsetting.

Around the age of twenty, I had this amazing new lucid dream experience where I drove a Ferrari styled racing car through all manner of hazards, causing damage to everything except me, on purpose without injury. I jumped off of cliffs, rammed into semi tractor-trailer rigs, and anything else I could find, and just kept on going. Nothing stopped me.

The experience was incredibly exhilarating and I woke up in a sweat. I felt like I conquered the world. I went back to sleep and continued the dream. I say it was lucid because I seemed to have control in the dream. I was able to make decisions within the dream, at least that's my perception. I also made some decisions while awake and took them back into the dream. Periodically, I revisited the dream, but seemed to have lost interest in it within a year.

In the early 1980s, I realized that I used lucid dreams for practical purposes, and that they would generally occur when I saturated my thoughts with anything for a season. If I were studying calculus, I had calculus adventures. When I built my studio, I had architect adventures. I literally learned to go to sleep with a problem and wake up with the answer to the problem, a solution that I could implement in the real world.

I continually use lucid dreams to develop computer software, or advance any project that I happen to be working on. Lucid dreams quite enjoyable, sort of like a video game, and as far as I can tell, anyone can develop the skill. The initial key is to realize, in your dream, while you are dreaming, that you are dreaming. After that, you need to be faithful to write down whatever you wake up with, so it doesn't evaporate. Then, you will find the real world benefit from them, and naturally get better at it because you experience the reward.

My personal view about dreaming is that it's related to your heart having an opportunity to file away your new beliefs, and as your subconscious mind, it has unfettered opportunity to address your current burdens. I suspect meditation techniques can also put you in touch with your heart. While you sleep, your conscious mind rests, yielding to your subconscious mind, your heart.

While I'm at it, the world has referred to these dimensions collectively as the mind, subconscious, conscious, and super conscious, which is the heart, mind, and spirit. Strictly speaking, unconscious mind is your unseen parts that are not in your conscious mind, and those are the dimensions below and above your mind, which are the subconscious and super conscious, or heart and spirit. In reality, they are not part of your mind; they have their own unique perspective, and their own names.

What's going on during a lucid dream? Here's what my model tells me. Dreaming is primarily a heart activity, while your conscious mind is sleeping. However, a lucid dream connects the two, but the heart is taking the lead, and the mind is cooperating to a smaller degree. You are still sleeping, but in a slightly different mode. Your mind can now ask questions, and make decisions to change the focus of your heart, to guide your heart in its exploration, and still yield to the heart to a significant degree. In fact, I sometimes wake during a lucid dream because my mind takes complete control. It's frequently disappointing to realize that I am no longer flying; flying was great fun while my conscious mind yielded to my mind, in my sleep, I believed I was flying while my heart was in control.

1 Kings 3:5

In Gibeon the LORD appeared to Solomon in a dream by night: and God said, Ask what I shall give thee.

If god can give you dreams, how does that happen? Here's what my model tells me. God communicates silently to you from his spirit to your spirit. When God appeared to Solomon, Solomon became aware of it in his conscious mind. Solomon was dreaming while he slept, and yet his mind was aware, which could be lucid dreaming, plus the communication in his spirit. All of the unseen parts of Solomon were active in his dreaming.

57 Prophecy

There are two kinds of prophecy, foretelling and speaking forth to bring about by faith. It just might be that they both align to become one when you consider that the very act of the foretelling is the calling forth, by faith, that which doesn't exist into existence, exactly the same process by which god called forth creation by his words; god could be using you to call forth things that do not exist to serve his purpose; your faith can be used by god to change things on this earth.

A very strange thing about all of my experiences, with faith changing things, is that they occur without forethought, as though I was prompted. I don't think I want to tell you about any of them. However, I did tell you about another place that the same dynamic was involved. I can see that moment in my life, in my mind so vividly right now. When I was playing drums at church and said "Lord heal it" and it was healed. I have never before, and never since, presumed to command the Lord in any way. As I reflected on it shortly afterwards, it one bothered me enough to repent from it. There was zero forethought when those words came flying out of my mouth. I have no idea why I would even consider saying such a thing, it doesn't align with any theology or teaching that I have ever heard; it puzzled me at the time. I now see the parallel that all of my calling forth by faith moments, which have changed things, had no forethought on my behalf; they seemingly escape my mouth without my conscious decision. Yet, they do not all seem infallible or redeemed; they can contain my selfishness in them, which means that there's some element that they come from me. I told you of the one that was selfish, god would not cause me to say such a thing, unless he had a specific lesson in mind, and their purpose isn't always clear. Those aren't the only puzzling cases in my experience. This verse talks about being told what to speak, but I'm not completely convinced it refers to immediately before one speaks; in that same hour isn't necessarily in that very minute.

Matthew 10:19

But when they deliver you up, take no thought how or what ye shall speak: for it shall be given you in that same hour what ye shall speak. For it is not ye that speak, but the Spirit of your Father which speaketh in you.

58 An Inconvenient Truth

This definition of a dimension cannot be improved upon, and can be used to challenge any model of everything. The bottom line is this, "unique perspective" specifically eliminates all possibilities of more than one dimension having the same scope as another, and therefore there cannot be any parallel universes or any such nonsense because unique means unique. If it's not unique, then what is your definition of a dimension? Give me a better definition of dimension, or you don't have a leg to stand on. Show me the math, and show me the definition; without those you are making up a fairy tale. Don't give me any of that nonsense about "I see the possibility for it in the math" without defining dimension because it means nothing.

For example, negative numbers are extremely useful, but they don't define reality. Reality is one fish, or no fishes, but not negative one fish. What is a minus one fish? I tell you what it's not; it's not a definition of reality.

The same goes for infinite dimensions. I touched on it before, but if a dimension is unique, it has a purpose and a scope, it has uniquely defined reason for being, and there in not an infinite need for definitions. Long before you near infinity, the need for places to put things uniquely is long since satisfied. In fact the number is surprisingly small. I say it is twelve.

Just because you see possibilities in the math, does not mean they are possible definitions for reality. Math allows for infinite dimensions, but if there is a definition for a finite existence, there is no room for infinite dimensions. If the definition of entity is that which exists, then it has a finite definition, it is something right now, and that something has a finite definition; an entity must exist in an existence.

If the definition of dimension is "unique perspective", then that implies that everything, which can be named, has a place within the model of everything; that is what makes it a model, or theory, of everything. If you can't explain that, then you aren't there. I can explain that because I'm there.

I can tell you insights about almost any subject, having no prior knowledge of the subject. Put me to the test, tell me enough about any topic that you think you know more than me, tell me enough to have some understanding of what you believe, and I can tell you which parts are true and which parts are false and what direction the field of study is going to advance in next, and what is preventing it from advancing. Is that practical enough? Hey, try me. I might have forgotten more about our world than any scientist that ever lived has ever discovered. It really bothers me to say things like that.

That's an appalling statement. How could anyone say that with a straight face? If it is true, why have you not said anything before now? There are many reasons. Trust me, I have asked myself that question so many times it haunts me.

I'm amused by watching disciplines wander all over the place, knowing they are wasting time looking in the wrong place because they don't even know what the questions are. You can't find the answer to anything significant without knowing enough questions to cause you to explore in a fruitful direction. You might be exploring one question and discover a fact about a different question, but you don't get anywhere without knowing a real question.

The things science continues to ask are the wrong questions, and as long as they continue to do that, I know that they will not discover what I know until I decided to tell them, or when God decides to give it to someone else to tell them. I've watched them languish for decades while I was confident that I could wait forever and they wouldn't figure it out. Every now and then they gain a tiny glimmer of advancement in a productive direction and I know why.

The funny thing about truth is... it's true. It doesn't matter where it comes from, though I can tell you that the source of all that is good and true is god. You don't need religion to tell you what is true, but neither can you deny the truths that religion delivers, and if it delivers, then you are a fool to ignore it; throw the baby out with the bathwater and enjoy a future filled with regret and despair, the choice is completely yours, just like you have the choice to pour a gallon of gasoline on your head and light a match; you wouldn't be the first, and you won't be the last. No matter your choice, truth will always be... well, true.

59 Life is All About the Heart

If your heart is full of god's wisdom, then it serves you well to listen to it, but the heart that is not filled with god's wisdom is folly.

Proverbs 16:21

The wise in heart shall be called prudent: and the sweetness of the lips increaseth learning. Understanding is a wellspring of life unto him that hath it: but the instruction of fools is folly. The heart of the wise teacheth his mouth, and addeth learning to his lips. Pleasant words are as an honeycomb, sweet to the soul, and health to the bones.

For a heart that is not filled with the wisdom of god, the worst advice anyone can give another person is to tell them to follow their heart. Girls are following their heart when they pick bad boys and get used and dumped. Pedi files are following their heart when they abuse children. In fact, I think we can say that all criminals are following their heart when they commit crimes. I'm very cautious about using the word 'all', but the preponderance stands.

You must not follow your heart in order to develop as a person, in every case. The only way to develop your personality is to make a conscious decision not to follow your heart. If you only followed your heart, you would forever behave as an infant, crying because you were hungry and every other discomfort that you encountered. The only reason that you don't still act like that is because you didn't follow your heart, many, many times, over and over, again.

Here's the deal... your heart has exactly one job, to remind you of what you believe, what you already believe, it's old news, and it can be trusted to remind you of old news. What it is completely incapable of is telling you what you should believe. All changes to your belief system occur in your mind as a result of evaluating new perspective. Then, when you change your mind about what you believe, your heart holds onto that new belief, and whenever you need to be reminded of it, poof, there it is. Your heart is dependable, loyal, faithful, and trustworthy, but only to remind you of your latest judgment on any topic.

So, if your latest judgment on being attracted to bad boy was 'yes, i want that', then your heart will be faithful to deliver it to your mind when you see a bad boy. It might also remind you, that you considered that picking the bad boy is risky, in fact it has always ended in heartache, but it will remind you that your desire was victorious over your caution, and if you follow your heart, you'll pick that bad boy again, be abused again, be dumped again, and live with the heartache again. That bad boy knows it and exploits it; you're so predictable that this hunting strategy is employed by billions of bad boys, and there is a huge cost to civilization because of it.

In order to develop your personality, to become a complex and productive person, as a member of a family and a society, you must incorporate millions of decisions where you choose not to follow your heart, and in fact going directly against, contrary to, the thoughts of your heart.

Life, itself, is a never-ending series of forks in the road, where you decide to follow, or not to follow, your heart. You will either reinforce an existing belief, or you will record a new belief. As you grow, your belief system becomes more complex, and if you have character, your belief system will become more solidly entrenched in a consistent direction, based upon principles. Without principles, you are nothing more than an infant, blown about to and fro, or worse.

Following your heart is the same thing as doing whatever you want, following your desire. Is it productive to just do whatever you want? Do we say yes to criminals? Do we say yes to Hitler, and Stalin? Yes, Hitler, it's ok to kill millions of Jews because you want to. Do those words come from your lips? It's not ok on that scale, and it's not ok on a personal scale. In fact, it's not ok, as a principle, on any scale.

It is ok to choose certain liberties in your own life, to explore and grow, to find out who you are, to make mistakes, but there are limits. You should be setting personal limits, beyond which you believe you should not cross. Only then, can you begin to follow your heart and have a good result. Only a principled person can depend on their heart without risking regretful outcomes.

The purpose of the heart is to allow you to live a more complex life. If you fill it with crap, you can only trust it to send you down the path to ruin. If you fill it with truth, you can respond in complex situations accurately without having to weigh all of the possible outcomes.

You use your heart to select something to eat for lunch. What if you had no memory of what you had eaten before? Now you become hungry, and you look around at the trees and the dirt and wonder? Is that edible? Will it satisfy my hunger? You see how quickly your life disintegrates without the ability to depend upon your heart. If you allow your heart to have its way with you all the time, you will only have ice cream for dinner, and you will die prematurely.

If you intend to mature, to grow up, then you must act against the beliefs of your heart that do not conform to the wisdom of god. There is no other way. The bigger issue is this: God is going to look upon your heart to see what you stored in there, and it counts more than anything else.

60 Hidden Things

This model was a hidden thing, I called upon the lord and he answered.

Jeremiah 33:2

Thus says the LORD who made the earth, the LORD who formed it and established it, the LORD is His name: Call to Me, and I will answer and show you great and unsearchable things you do not know.

61 The End From the Beginning

Can we know what to expect from the future? Yes, because the bible tells us so. It is the only book, ever written, that accurately tells the end from the beginning. Throughout history, it revealed god's plan, along with many details that have come to pass, exactly as foretold. It is the only book in that category because it was inspired by the one true god, who knows the end from the beginning.

Was Jesus resurrected? Will you be resurrected? The bible says so. Have you thought about eternity? It's a daunting topic to wrap one's mind around. What does the bible say? We shall live forever. The question then becomes... where shall we live?

62 The Dividing Line

Who needs to be saved? "All have sinned and come short of the glory of God." Why do we need to be saved? "Without the shedding of blood there is no forgiveness of sins." From what do we need to be saved? The wages of sin is death. God deals with us justly and contractually. If God is not just, then there is no meaning of life. God must not lie, for if he lies, then everything falls apart. Every word that God says must return to him exactly perfect, or he is not God. Every scripture must be perfectly fulfilled, or there is no truth, and the Bible is useless. This life, which you are living now, is but a blip in time, and you shall live beyond your grave. The question then becomes, what sort of life shall you live, for all of eternity, after your body dies? "Rid yourselves of all the offenses you have committed, and get a new heart and a new spirit. Will you die?"

63 Once Saved Always Saved

Romans 10:9

That if thou shalt confess with thy mouth the Lord Jesus, and shalt believe in thine heart that God hath raised him from the dead, thou shalt be saved. For with the heart man believeth unto righteousness; and with the mouth confession is made unto salvation.

I love how it supports my model when it says that you believe in your heart, and confess with your mouth, which is controlled by your conscious mind.

John 6:35

And Jesus said unto them, I am the bread of life: he that cometh to me shall never hunger; and he that believeth on me shall never thirst. But I said unto you, That ye also have seen me, and believe not. All that the Father giveth me shall come to me; and him that cometh to me I will in no wise cast out. For I came down from heaven, not to do mine own will, but the will of him that sent me. this is the Father's will which hath sent me, that of all which he hath given me I should lose nothing, but should raise it up again at the last day. And this is the will of him that sent me, that every one which seeth the Son, and believeth on him, may have everlasting life: and I will raise him up at the last day.

By faith, we believe that his sacrifice on the cross is enough to save our immortal soul from being separated from him forever. By repenting from our sin, and confessing with our mouth, we receive justification before god, and from that point on, God's mercy looks at us, just as if we had never sinned. Your salvation is a settled issue, and that is the beginning of your Christian walk, the first part of his promise to you, in the new covenant, but that mercy is only half of the promise. To escape judgment in the tribulation, you also need sanctification, the other part of his promise.

We must be restored to god because we have sinned, and the way back to him is through Justification and Sanctification. We are justified by faith when we believe in the sufficient sacrifice of Jesus on the cross to pay the price for our sin. That is the first step in our restoration, but we cannot live in the presence of god without also being sanctified.

I'm trying really hard not to include too may bible verses in this book. I wanted it to be mostly my expression of the model. I spend a great deal of effort proving my logic on this topic and the only way to do that is allow the bible to tell us what it says, and for that reason, I'm not going to give you the whole story here, but I'll tell you where you can see it, but before I do, I need to tell you something important.

64 With Whom Did God Make Covenants?

With whom did god make the old and new covenants? If you are in the new covenant, you are Israel. God made the old and new covenants with Israel, not with Christians. The new covenant, to be saved by faith in the blood of Jesus, was made with Israel. The new covenant with Israel writes the same Old Testament laws on the hearts of God's people in the New Testament. If you are in covenant with god, you are Israel; if you have been grafted into the tree, you are part of the tree, and everything he said about the tree is about you, Israel. The new covenant is a different covenant with the same Old Testament law, God's law that has not changed, but is now written in the hearts of God's people.

Jer 31:31

Behold, the days come, saith the LORD, that I will make a new covenant with the house of Israel, and with the house of Judah: Not according to the covenant that I made with their fathers in the day that I took them by the hand to bring them out of the land of Egypt; which my covenant they brake, although I was an husband unto them, saith the LORD: But this shall be the covenant that I will make with the house of Israel; After those days, saith the LORD, I will put my law in their inward parts, and write it in their hearts; and will be their God, and they shall be my people. And they shall teach no more every man his neighbour, and every man his brother, saying, Know the LORD: for they shall all know me, from the least of them unto the greatest of them, saith the LORD: for I will forgive their iniquity, and I will remember their sin no more.

God's law, he calls it "my law", has not changed, it cannot change, he cannot change, he is bound by his word, which will never return void.

Matt 5:17

"Do not think that I came to destroy the Law or the Prophets. I did not come to destroy but to fulfill. For assuredly, I say to you, till heaven and earth pass away, one jot or one tittle will by no means pass from the law till all is fulfilled. Whoever therefore breaks one of the least of these commandments, and teaches men so, shall be called least in the kingdom of heaven; but whoever does and teaches them, he shall be called great in the kingdom of heaven. For I say to you, that unless your righteousness exceeds the righteousness of the scribes and Pharisees, you will by no means enter the kingdom of heaven.

Again, from the mouth of your lord, Jesus said, "till heaven and earth pass away, one jot or one tittle will by no means pass from the law till all is fulfilled." This isn't a veiled reference and it's not ambiguous; they are the words of Jesus quoted in the New Testament. Look at the words of Jesus in the New Testament about the importance of the law; he lists it first, weighted by importance in the transfer of knowledge.

Matt 23:23

Woe unto you, scribes and Pharisees, hypocrites! for ye pay tithe of mint and anise and cummin, and have omitted the weightier matters of the law, judgment, mercy, and faith: these ought ye to have done, and not to leave the other undone. Ye blind guides, which strain at a gnat, and swallow a camel.

65 Hot Off the Press

Ez 20:12

Moreover also I gave them my sabbaths, to be a sign between me and them, that they might know that I am the LORD that sanctify them.

Ez 20:19

I am the LORD your God; walk in my statutes, and keep my judgments, and do them; And hallow my sabbaths; and they shall be a sign between me and you, that ye may know that I am the LORD your God.

Matt 5:17

Think not that I am come to destroy the law, or the prophets: I am not come to destroy, but to fulfil. For verily I say unto you, Till heaven and earth pass, one jot or one tittle shall in no wise pass from the law, till all be fulfilled. Whosoever therefore shall break one of these least commandments, and shall teach men so, he shall be called the least in the kingdom of heaven: but whosoever shall do and teach them, the same shall be called great in the kingdom of heaven.

Ecc 12:13

Let us hear the conclusion of the whole matter: Fear God, and keep his commandments: for this is the whole duty of man. For God shall bring every work into judgment, with every secret thing, whether it be good, or whether it be evil.

1 John 5:1

Whosoever believeth that Jesus is the Christ is born of God: and every one that loveth him that begat loveth him also that is begotten of him. By this we know that we love the children of God, when we love God, and keep his commandments. For this is the love of God, that we keep his commandments: and his commandments are not grievous.

1 John 2:3

And hereby we do know that we know him, if we keep his commandments. He that saith, I know him, and keepeth not his commandments, is a liar, and the truth is not in him. But whoso keepeth his word, in him verily is the love of God perfected: hereby know we that we are in him. He that saith he abideth in him ought himself also so to walk, even as he walked.

We're told the terms of the contract is to keep the law; to believe in Jesus to be born again, and demonstrate that the love of god is in your life, by keeping the law; to keep his commandments in the new testament; and demonstrate that you know him by keeping the law, as Jesus did.

This is the short version of sanctification, no matter when or how the process occurs; god cannot live with you until your soul has been completely sanctified. For all who have died in him, god has promised to complete their sanctification. However, we are living in a time that we need a special covering to protect us while we are alive on earth, and the promise that god makes to protect us depends on us entering into that promise, by covenant.

In the bible, you are Israel; god didn't make a covenant with Christians, he made covenants with Israel, and if you are in any covenant with god it can only mean that you have been grafted into Israel; the Ten Commandments shall never pass away, they apply to you right now; the fourth commandment requires you to observe the Sabbath as a sign that you entered into covenant with him; that you trust him to sanctify you, which is the opposite of the old covenant, where we promised we would do things; the new covenant is god promising us that he is able to sanctify us; there are many quotes from the bible where Jesus said to keep these commandments, you must be sanctified to live with him; your sanctification depends on you entering into a covenant with him; the lynch pin is the sign between god and man, the Sabbath; no matter how badly you fail, god is able to complete your sanctification, but you must observe the Sabbath to get in for it to provide safety in this time of tribulation that is coming; all of god's covenants have included the two promises of mercy and restoration, which are justification and sanctification; restoration is complete after justification and sanctification, depicted by the pattern in the sanctuary. This paragraph, about the Sabbath and sanctification, is the only topic in this book, which is hot-off-the-press, in my life in December of 2021, it brought me to my knees, and you can read the whole story for free at foreshown.com

66 Belief and Unbelief

To be justified, your responsibility is to believe the gospel, this gospel…

You must believe that you need a savior because you are part of fallen man, all have sinned and deserve death for it, that Jesus is God, and he lived perfectly on Earth so that he could be a pure sacrifice to pay your sin debt, and he paid your debt in full when he died for your sins, and he walked out of the tomb on the third day after the cross, defeating death itself.

When you believe that gospel, you are justified, which is the beginning of restoration. To trust is to believe, and to believe is to trust; they are the same, when you believe it, you count on it, you depend upon it, you live by it, and you trust that it is so. Belief in truth produces faith, and faith is powerful. Without belief in truth, there can be no faith. Without faith it is impossible to please God. Without belief, it is impossible to be saved. You were made to believe. Your very construction is built to believe. You were designed for the purpose of believing. You cannot accomplish anything without believing. You learn because you believe. You learn when you believe. The act of learning is taking in new revelation until you can believe. There is no partial believing, either you believe and it is so, or, you do not believe and, for you, it is not so. Either, you believe or you do not believe; in between there is only a process of soul searching and continued revelation, or darkness. During the whole process of understanding something new, you do not believe until the moment arrives when you believe. You can believe a lie and you can believe the truth. What you believe determines what you are; you are what you believe. What you believe, now, determines your path in both, this life and in eternity. You can have a successful career, in this life, when you believe enough of the accepted truth about a field of study to become useful, for example. How can you be of any use, to any one, in an area that you do not understand? Shall we begin a new career path for those who know nothing? Who will pay us for that?

Your spirit is immediately redeemed when you are justified; God put his Spirit in you. Your soul, your mind and heart, is in the process of being redeemed, by the renewing that comes from the decisions you make with your mind when you believe revealed truth, and hide them in your heart, while you live on this earth before your body dies, which is sanctification. Your body is redeemed when you receive your new body, a glorified body not made with hands, after you physically die. You are then fully equipped to live the same unseen world that god lives in, regardless of which territory in that world becomes your final destination. If you are sanctified, which as far as I know means that you entered into the covenant with god by observing the Sabbath, and trusted that he is able to sanctify you, then you live in heaven with god. If you are not sanctified, then you spend eternity apart from god. That is my best revelation on that topic to date.

In eternity, you will know as you are known. Does that sound like a process? It doesn't say you will come to know as others come to know you. What if the dividing line is physical death? What if your soul is immediately and completely redeemed, in the same instant that you receive your glorified body? That means that you will not spend eternity learning and growing in Christ, but more than that...it means that your soul is set forever, everything that you will ever believe, you already believe, and God cannot save you after that. God cannot save you after you die. Does God change? How much did God know, from the beginning? All. Does God believe something today that he did not believe yesterday? No. And after you are restored, neither will you.

When you are with God in eternity, you will be of one spirit, the Holy Spirit, and you shall have the mind of Christ, able to rightly divide the word of truth. Your heart will contain your individuality; your heart is your personality, it defines who you are. The goal is to have the heart of the Father. "I will give you a new heart and put a new spirit within you; I will remove your heart of stone and give you a heart of flesh." That is God's job; that is God's responsibility. He gave his word that he will do this for you, but does it happen when you believe the gospel, or does it happen when you obey the commandments? Go find out, it matters.

After you are justified, you have the opportunity to continue the process of restoration and redeem your heart in sanctification, where you cooperate with god; the flesh means it is soft and easily changed, not hard and resistant to change. You have a window, between the time you are justified and the time your body dies, to make your heart as much like the heart of God, the Father, as you can; this is a life goal. The goal of life is to determine where you spend eternity; in the presence of God, or not. The meaning for your life is defined by how you execute what God intended for you to do, God's purpose for your life, your role in his plan. The more meaningful your life, and the greater your heart is like the heart of the Father, and the greater your reward in eternity. While we are waiting for the adoption, and the redemption of our body, we have time to fulfill his purpose for our lives. Only God knows the end from the beginning, and he shares revelation to each of us in varying degrees.

The secret things belong unto the LORD our God: but those things which are revealed belong unto us and to our children for ever, that we may do all the words of this law.

We all have the Bible, which is the revealed word of God, but Jesus reveals his secrets to anyone he chooses because they belong to him. After he reveals them to us, they belong to us; he expects us to believe them, to believe his word; he expects us to believe him. Jesus came from heaven, to earth, to meet you at the cross, which is justification by faith, and he shows us the way back to heaven with him, according to the pattern in the sanctuary, which culminates with sanctification, according to the law.

67 ONE GOD

Is there more than one God? No. "For there is one God and one mediator between God and mankind, the man Christ Jesus." There is only one God. "I am the LORD your God, who brought you out of the land of Egypt, out of the house of slavery. You shall have no other gods before Me." In this text, gods, with a small g, implies that you have the power to allow things to be more important to you than God, and he's not having it, but that doesn't mean that there is more than one God. There is only one God, and all things, which were created, were created by him. "All things were made by him; and without him was not anything made that was made." "For in Him all things were created, things in heaven and on earth, visible and invisible, whether thrones or dominions or rulers or authorities. All things were created through Him and for Him." There is only one God. God revealed his character to us in the description of three persons so we could easily understand it.

"But now the righteousness of God without the law is manifested, being witnessed by the law and the prophets; Even the righteousness of God which is by faith of Jesus Christ unto all and upon all them that believe: for there is no difference: For all have sinned, and come short of the glory of God; Being justified freely by his grace through the redemption that is in Christ Jesus: Whom God hath set forth to be a propitiation through faith in his blood, to declare his righteousness for the remission of sins that are past, through the forbearance of God; To declare, I say, at this time his righteousness: that he might be just, and the justifier of him which believeth in Jesus."

"But if Christ is in you, your body is dead because of sin, yet your spirit is alive because of righteousness. And if the Spirit of Him who raised Jesus from the dead lives in you, He who raised Christ Jesus from the dead will also give life to your mortal bodies through His Spirit, who dwells in you."

We are adopted by God when we believe that Jesus is God, he lived without sin in order to qualify as a worthy sacrifice, he died to pay the price that the contract demanded for our sins, and he rose from the dead so that we can live with him as members of his family.

If the Holy Spirit, who raised Jesus from the dead, lives in you, then he will also raise you from the dead to live an eternal life with him. That's the promise. My question to you is what is your response? What do you believe?

DAYS OF CREATION

Genesis 1:1

In the beginning God created the heaven and the earth. And the earth was without form, and void; and darkness was upon the face of the deep. And the Spirit of God moved upon the face of the waters.

The First Day: Light

And God said, Let there be light: and there was light. And God saw the light, that it was good: and God divided the light from the darkness. And God called the light Day, and the darkness he called Night. And the evening and the morning were the first day.

The Second Day: Firmament

And God said, Let there be a firmament in the midst of the waters, and let it divide the waters from the waters. And God made the firmament, and divided the waters which were under the firmament from the waters which were above the firmament: and it was so. And God called the firmament Heaven. And the evening and the morning were the second day.

The Third Day: Dry Ground

And God said, Let the waters under the heaven be gathered together unto one place, and let the dry land appear: and it was so. And God called the dry land Earth; and the gathering together of the waters called he Seas: and God saw that it was good. And God said, Let the earth bring forth grass, the herb yielding seed, and the fruit tree yielding fruit after his kind, whose seed is in itself, upon the earth: and it was so. And the earth brought forth grass, and herb yielding seed after his kind, and the tree yielding fruit, whose seed was in itself, after his kind: and God saw that it was good. And the evening and the morning were the third day.

The Fourth Day: Sun, Moon, Stars

And God said, Let there be lights in the firmament of the heaven to divide the day from the night; and let them be for signs, and for seasons, and for days, and years: And let them be for lights in the firmament of the heaven to give light upon the earth: and it was so. And God made two great lights; the greater light to rule the day, and the lesser light to rule the night: he made the stars also. And God set them in the firmament of the heaven to give light upon the earth, And to rule over the day and over the night, and to divide the light from the darkness: and God saw that it was good. And the evening and the morning were the fourth day.

The Fifth Day: Fish and Birds

And God said, Let the waters bring forth abundantly the moving creature that hath life, and fowl that may fly above the earth in the open firmament of heaven. And God created great whales, and every living creature that moveth, which the waters brought forth abundantly, after their kind, and every winged fowl after his kind: and God saw that it was good. And God blessed them, saying, Be fruitful, and multiply, and fill the waters in the seas, and let fowl multiply in the earth. And the evening and the morning were the fifth day.

The Sixth Day: Creatures on Land

And God said, Let the earth bring forth the living creature after his kind, cattle, and creeping thing, and beast of the earth after his kind: and it was so. And God made the beast of the earth after his kind, and cattle after their kind, and every thing that creepeth upon the earth after his kind: and God saw that it was good.

And God said, Let us make man in our image, after our likeness: and let them have dominion over the fish of the sea, and over the fowl of the air, and over the cattle, and over all the earth, and over every creeping thing that creepeth upon the earth. So God created man in his own image, in the image of God created he him; male and female created he them. And God blessed them, and God said unto them, Be fruitful, and multiply, and replenish the earth, and subdue it: and have dominion over the fish of the sea, and over the fowl of the air, and over every living thing that moveth upon the earth. And God said, Behold, I have given you every herb bearing seed, which is upon the face of all the earth, and every tree, in the which is the fruit of a tree yielding seed; to you it shall be for meat. And to every beast of the earth, and to every fowl of the air, and to every thing that creepeth upon the earth, wherein there is life, I have given every green herb for meat: and it was so. And God saw every thing that he had made, and, behold, it was very good. And the evening and the morning were the sixth day.

69 Science Says

I just searched for these answers on the web, don't take stock in it, I only wanted samples for entertainment. The answers matter not.

The First Day: Light - the universe is 13.8 billion years old

The Second Day: Firmament - the earth 4.543 billion years

The Third Day: Dry Ground - 299 million years ago

The Fourth Day: Sun - the sun is 4.603 billion years old, Moon 4.5 billion years ago, the first Star 100 million years after the big bang

The Fifth Day: Fish - 530 million years ago   and Birds originated 60 million years ago

The Sixth Day: Creatures - on Land 360 million years ago

There is no reason to chart those dates for the purpose in trying to determine the definition of God's day to make it fit. I couldn't possibly care what science has to say about the timing of anything. To think that Cosmologists are ever going to come up with the theory of everything is preposterous. The finches on Galapagos are more likely to populate the earth with humans after total annihilation from global nuclear war than cosmologists arriving at the accurate theory of everything.

Your dark matter and dark energy do not live in the physical universe, and therefore shall never be discovered by the exploration of the physical universe. Look all you want, there is plenty to discover, but the theory of everything is not there because the universe is not everything. Neither is the possibility of a multiverse. I can kill all of your silly theories in one paragraph. Let's change the world, now, and forever. Ready?

The short version is this, Einstein gave us the theory of relativity, and various, which proved that the physical dimensions are continuously linked to time, and the theory has proven to be true, experimentally verified by science. I am going to give you the remainder of that theory, the one that Einstein spent his whole life looking for, and never found. The theory of everything adds to Einstein's theory and defines that everything belongs in a single continuum of dimensions, and I can define those dimensions, and the ramifications includes this: nothing can ever travel to another physical dimension because none can exist. The simple fact, the key to the proof, is this, the definition of a dimension is unique perspective. Unique means that no other can exist, if it existed, it would not be unique. You cannot have it both ways and call it honest logic. Either, you have to define a dimension better than me, or you do not have a leg to stand on.

Are you looking for a theory that is short and sweet? Is that part of your criteria of truth? I could give a third grader the answer that they can understand, to a significant degree, in less than 5 minutes. Oh, I have the truth, and not because I am smart. I might be exaggerating when I say that I have read five books in my entire 62 years of life. I failed at college and dropped out. I do not have the truth because I'm smart; I have it because it was given to me.

So, which is it? I'm waiting. Prove me wrong. Do you even have an argument? Don't even contact me if you cannot define a dimension better than me because your words are meaningless imagination. You are looking in the wrong place; you don't even know what the questions are, much less the answers. I have watched you wallow in stupidity for nearly four decades, with confidence that you never had a chance to publish the answer before me. I am the authority on this subject because it was given to me, and it is my story to tell, whenever I choose to tell it. I wish that I could have known Einstein, or even Hawking. Hawking contributed a lot of excellent exploration, but he was a fool. He said things that he had no basis for. That's how science works, you say crazy stuff and then you try to prove it. Thank you Steven, for all you did; too bad a bunch of it was trash. I wish I had the opportunity to say it to your face, my bad, I could have, but you are not my priority, RIP.

Here's the deal, pickle, God creates by speaking reality into being; everything which did not exist, God created with words. That does not means that the day he said it, that it instantly occurred and became physical reality. God's part was the speaking, and could have spoken the words that caused the creation process for everything in the universe on 6 literal days, or even in 6 literal minutes, and that does not preclude him from creating the universe that required 14 billion years for you to see it. The very minute that the process became physical, the process was subject to physical laws, and physics requires time; time itself was born. God did his part in 6 literal days. He said so and I believe it. I am a believer, exactly as I was created to be, created by my creator, on one of those creation days. God did it all, perfectly fit together for all time, and beyond time. Only god, who lived outside of the universe, could have done that.

Is this ringing true to anyone? Poke a hole in it, take your best shot. I'm always amused that people think they are smarter than god. God laughs the laugh of derision, and I laugh with him because I know him and I believe him. Do you know how you can know that I know him? You know because you see me try to keep the law of god. I fail, I fail often, and I fail monumentally, but I declare to god and to the world that I believe his law is in effect and I observe the Sabbath because I trust that god is faithful to sanctify me so that I can eventually be fully restored to god. There is no other criterion that matters because that is what god said, and that is why the fear of god is the only indicator, by which I judge my fellow man. All I have to know about you, to accurately size you up, is to see how you fear god.

Matthew 10:28

And fear not them which kill the body, but are not able to kill the soul: but rather fear him which is able to destroy both soul and body in hell.

Derision is the use of ridicule or scorn to show contempt. Believers are mocked by unbelievers, here on earth, and strangely it goes both ways, god mocks unbelievers.

Psalm 59:8

But thou, O LORD, shalt laugh at them; thou shalt have all the heathen in derision.

Psalm 2:1

Why do the heathen rage, and the people imagine a vain thing? The kings of the earth set themselves, and the rulers take counsel together, against the LORD, and against his anointed, saying, Let us break their bands asunder, and cast away their cords from us. He that sitteth in the heavens shall laugh: the Lord shall have them in derision. Then shall he speak unto them in his wrath, and vex them in his sore displeasure.

Who wants a taste of god's wrath? Lucky you, it's coming, and coming soon. If you're reading this, and you don't die soon, you will get a front row seat, and shall be vexed in his sore displeasure. How do hundred pound hail stones sound to you? Is that sore enough? Get your house in order now; this is your last warning before the door closes.

70 Relationship is Everything

The following statement describes the process where you lose access to our physical universe, and gain full access to the unseen world.

Philippians 3:20

For our conversation is in heaven; from whence also we look for the Saviour, the Lord Jesus Christ: Who shall change our vile body, that it may be fashioned like unto his glorious body, according to the working whereby he is able even to subdue all things unto himself.

If you do not have a relationship with god through Jesus, what do you have?

71 Believe

Let's look at the process of believing. We already talked about how your mind makes decisions that you store in your heart as your beliefs. Let's look at the decision part of the process, which occurs in your mind. The decision occurs in your mind, but you don't get to arbitrarily choose what you believe. You believe what you believe. You can't fake it and you can't talk yourself into it. You believe it or you don't believe it.

It isn't a belief until it is. Before you believe, it can be a topic of interest, a speculation, a possibility, an interesting thought, a wish, a desire, or a dream. Free will gives you the ability to choose how you will respond, by using your beliefs as your frame of reference, from which you make a choice of how you respond. After you are convinced that it is true, then you believe it is true. You can be convinced that you want it to be true, even while you do not believe it to be true. You can be exposed to truth, and hear it, and comprehend it, and still not believe it. You can even believe that it is the truth and still not believe it. You don't believe it until your heart is open to believing it, and opening your heart isn't always done by you; the Lord also opens your heart to believe what he reveals to you.

You can believe the Bible is the truth and still not believe the Bible. You can believe that the Bible contains God's word and still not believe God's word. The flip side is also true; you can believe it, even when you don't want to believe it. You can believe it and not want it to be true. It can be true, whether you believe it, or whether you don't believe it. You can believe it, whether it is true, or whether it is not true. You can open your heart to believe lies and to believe truth. You can believe lies when you are deceived, and we are all deceived to varying degrees. It is this mixture of deceits and truths, which fill our heart, that form our belief system, upon which we make our choices, but all beliefs are not choices.

Much of what you believe, you acquire by choice, building precept upon precept, whether accurately, or not, but not all; some beliefs are not chosen by you. In an extreme case, for example, you slip off the edge of a cliff, and you immediately believe that you are falling to your death. You might or might not die, still you believe, having no prior consideration, other than a general knowledge that falling from a great height could be hazardous to one's health, and indeed physical life. Long before you hit the ground to find out the outcome, you have already believed.

Paul's experience on the road to Damascus caused Paul to believe God, without prior consideration. Even in his immediate response, "Who are you, Lord?" expresses his inability to comprehend, while immediately knowing. Without Paul's permission, Paul believed. Whether Paul wanted to believe, or not, Paul believed. Paul didn't open his heart to believe, the Lord opened Paul's heart to believe. Paul couldn't un-believe, Paul couldn't stop believing, Paul believed.

72 Hardening of the Heart

The only thing that hardens a heart is unbelief, and original sin is where it began. The heart is the essence of the personality, the core of one's being that contains one's beliefs. Like the physical heart, it pumps life, like blood through the body; it pumps beliefs through the unseen parts of a person. God looks upon the heart because the intents of the heart are the record of how much you believe God. In fallen man, the heart is evil above all things. David was a man after God's own heart; we should all be a people after God's own heart. That is the goal of life, all lives. The meaning of life, for each individual, is found in God's purpose for that life. When we are saved we instantly become citizens of heaven, our spirit is immediately and forever saved, destined to spend eternity in the presence of God. As a child of God, you can know that you are saved, but the body isn't going with you. The body you live in is corruptible, and it must die, but you shall receive a new body that shall never die; a body capable of carrying you through eternity in the presence of God, without pain and death. When is your body saved? It is saved when you are resurrected or transformed in the twinkling of an eye at the last trump. What then of the soul?

What is the soul? Have you ever slapped your fingers and palms together to make a noise? Did you call it clapping your fingers and palms? Of course you didn't, you called it clapping your hands. Fingers + palms = hands. In the same manner, the soul is a combination of things; your soul is your heart and your mind. Heart + mind = soul. If your spirit is immediately, completely saved when you believe that Jesus, the son of the living God, gave his life to reconcile you to God, was buried and resurrected, and then your body shall be resurrected at some later time, then when is your soul saved?

Your soul is continually being renewed by the choices that you make in your mind, all through your walk in this corruptible body, and shall be completely saved when you know as you are known, in the presence of God in heaven. Until then, each and every decision you make strengthens or weakens your heart; you decide whether you believe what God has said or you do not. Every choice brings you closer to, or farther from, God. You are a believing machine; you were created to believe. You were given the freedom to decide what you believe; free will is not a thing, it's not a part of man, but simply the state of not being forced to be a robot. You will decide where you spend eternity; you will decide whether you believe God, and you shall suffer the consequences for your decisions, and there will be many consequences, but none more important than where you live after this corruptible body dies after a very short time on Earth.

How do you store your treasures in heaven? Believe God now. Blessed is he that believes before he sees. A thousand times a day, you make significant choices; you continually decide what you believe, based upon your current state of revelation. As you walk through life, you become aware of the world around you, and your heart tells you what you currently believe about it, and your mind decides what you do next; each decision reinforces your beliefs or hardens your heart. You continually save what you believe, in your heart, and that is your treasure; it is who you are, it is the core of your being, it is your personality, and from the abundance of your heart, your mouth speaks, and your feet walk, and your hands touch. The state of current revelation implies one thing... a revealer. So, how does that happen?

Your heart is the collection of beliefs, and your mind is the gateway that determines what gets in there. No one can force you to believe, no one but God knows what's in your heart. You are to guard your heart. When something catches your attention, your heart pumps whatever you deemed important about that topic, in the past, and your mind becomes aware of your previous thoughts. Likewise, your spirit is tuned into the channel from your god, and your mind can receive thoughts that you must learn to discern. For the child of God, the Holy Spirit speaks to you in a still small voice, and you can hear the voice, right beside your current beliefs from your heart, and you must decide what is important. When you decide, your belief is stored in your heart, to be used next time you encounter a similar topic. The more you believe God, the more he can reveal to you. What do you seek?

73 SUMMARY

After Jesus gave us the Lord's Prayer in Luke's chapter 11:1, Jesus gave a stiff warning about misleading people.

Luke 11:52

Woe unto you, lawyers! for ye have taken away the key of knowledge: ye entered not in yourselves, and them that were entering in ye hindered.

I take that to heart to mean, woe to anyone who deceives God's children; this is what I am risking to tell you these things. If I deceive you, I will receive the reward in the woe. In Ezekiel, he warned false prophets:

Ezekiel 13:8

Therefore thus saith the Lord GOD; Because ye have spoken vanity, and seen lies, therefore, behold, I *am* against you, saith the Lord GOD. And mine hand shall be upon the prophets that see vanity, and that divine lies: they shall not be in the assembly of my people, neither shall they be written in the writing of the house of Israel, neither shall they enter into the land of Israel; and ye shall know that I *am* the Lord GOD. Because, even because they have seduced my people, saying, Peace; and *there was* no peace.

Again, in Ezekiel, a warning to women teaching magic:

Ezekiel 13:17

Likewise, thou son of man, set thy face against the daughters of thy people, which prophesy out of their own heart; and prophesy thou against them, And say, Thus saith the Lord GOD; Woe to the *women* that sew pillows to all armholes, and make kerchiefs upon the head of every stature to hunt souls! Will ye hunt the souls of my people, and will ye save the souls alive *that come* unto you?

Am I claiming "Thus sayeth the Lord" when I tell you that he gave this thing to me? I take these warnings very seriously. I fear nothing more than the Lord who can banish my soul to an eternity away from him. Again in Matthew:

The Dimensional Theory of Everything

Matthew 23:13

But woe unto you, scribes and Pharisees, hypocrites! for ye shut up the kingdom of heaven against men: for ye neither go in *yourselves*, neither suffer ye them that are entering to go in. Woe unto you, scribes and Pharisees, hypocrites! for ye devour widows' houses, and for a pretence make long prayer: therefore ye shall receive the greater damnation. Woe unto you, scribes and Pharisees, hypocrites! for ye compass sea and land to make one proselyte, and when he is made, ye make him twofold more the child of hell than yourselves.

I do not want to answer to god for anything he specifically said woe to; that sounds like a bad day to beat all bad days. Yet, I say these things to you because I fear only god, I believe these things, and without faith it is impossible to please god, so I step out in faith:

If i am a believing machine, how am i put together? In the end, you are responsible for the revelation that you have received. Do you believe that Peter walked on water? If you don't understand that question, why are you here? If you understand the question, but you believe that this story is some sort of allegory, then you heard nothing. If you don't believe that Peter literally walked on water, then you need to stay as far away from me as you can because the devil wants to steal from me, kill me, and destroy me, and he will do that to you, even if only as an attempt to get to me. You might not be a threat to him, but I am, and that makes me an active target. He doesn't just theoretically or generally want to kill me; he knows me, he fears Him who is in me, the enemy believes that he will reign longer if he can derail me from my purpose, and he will use you to get to me. You can become collateral damage in the war of good against evil by simply being near or dear to me.

What if these dimensions define absolute reality? The universe comes from nothing, empty, formless, and void. The dimensions are:

1) A unique perspective of the physical universe in which your natural body resides, height in the seen world.
2) A unique perspective of the physical universe in which your natural body resides, width in the seen world.
3) A unique perspective of the physical universe in which your natural body resides, breadth in the seen world.
4) Time, a unique perspective of the physical universe in which your natural body resides for a finite lifetime.
5) The collection of time in you, where you store who you are for all time, the sum total of your beliefs, your heart.

6) Where you gain perspective on time, change your beliefs, and decide who you are, your mind.
7) Where you receive perspective gain from beyond your senses, where you commune with God, what you really are, spirit.
8) A unique perspective of the spiritual universe in which your glorified body resides, height in the unseen world.
9) A unique perspective of the spiritual universe in which your glorified body resides, width in the unseen world.
10) A unique perspective of the spiritual universe in which your glorified body resides, breadth in the unseen world.
11) Eternity a unique perspective of the spiritual universe in which your glorified body resides for an infinite lifetime.
12) Maybe, just maybe, the Book of Life is a special dimension that gains perspective on eternity.

You can think of this this dimensional continuum as two worlds, the seen world and the unseen world. Better yet, 2 universes would probably describe them more accurately. I'm not the authority on the definition of universe, and don't pretend to be. The point is, these unique boundaries describe all that there is, or shall ever be until god destroys the seen world and recreates it. Each dimension gains unique perspective on the other, as one progresses through them. You could look at as the chart I presented, depicts how the supernatural overlaps the natural. That's a natural perspective, from a physical point of view, but we know that the supernatural world existed before the natural world. In reality, the natural and supernatural parallel dimensions of space and time dimensions exist together. God is here, the unseen universe is all around us, and it coexists with the seen universe. Its influence is felt in what science calls dark matter and dark energy. Everything came from somewhere; the big bang was the moment that the seen universe sprang from the unseen universe; the seen universe exists within the unseen universe, intermingled together separated in unique perspectives by strict definition of functionality and perspective gain.

A dimension is strictly defined as "unique perspective". Points, lines, planes ... explain unique perspective in the physical dimensions, culminating in the actual order, which follows the perspective gain pattern. Einstein's theories, which have been scientifically proven through experimentation in the physical universe, undeniably state that we exist in at least the four dimensions of space and time. That, alone, proves without any doubt, that we exist in multiple dimensions. We are multi-dimensional creatures. For all of us self-aware multi-dimensional creatures, the question then becomes: "In how many dimensions do we function?". How many dimensions are required to contain everything? What are they and what makes them unique? What are their purposes? If you have the correct model of reality, you can answer these questions. This structural definition, which I have presented, of all reality, answers these questions; this model provides insight which allows one to gain perspective on many things and answer many other questions; reality becomes less mysterious, understanding abounds, and life is simplified. Prove it, or disprove it, for yourself. You are a believing machine. You were designed to believe. You were created with the capacity to operate in these dimensions, the same dimensions in which God operates; you were created in His image. The seventh dimension is spirit; that's what makes us different from a caveman and all the rest of the animal kingdom. God made us in his image, and that means that we require seven dimensions to be complete; to function in the ways that God functions.

God told us to keep the Sabbath holy because it is spiritual, the day that signifies that we have a spirit and are able to commune with God, but more than that, he said it is a sign that we enter into a covenant with him where he promises to sanctify us, and we believe he is capable of accomplishing that, and we trust him to restore us to god, so that we can live in eternity with him, even though we currently are fallen in nature, and have no other way back to god.

You live in the same two universes that He lives in, and there are no more universes because there is no need for any; none can be defined according to the definition of dimension, which is unique perspective.

Jesus said, "My kingdom is not of this world. If it were, my servants would fight to prevent my arrest by the Jewish leaders. But now my kingdom is from another place."

Jesus didn't walk to Area 51 to catch a shuttle to his planet; He was talking about the unseen world. There is a second universe, but not a third universe. There cannot be a multi-verse because there is nothing left to define; I challenge you to name anything that doesn't have a place in this model. Anything that can find a place in this model cannot live in another dimension because that would break the definition of unique perspective, and therefore dimension, itself.

Granted, some things exist across more than one of these dimensions. For example, your physical body (or any physical body) exists in 3 dimensional space and time. That doesn't break the definition of dimension. For your physical body to exist in some parallel 3 dimensional space and time would break the definition of uniqueness; those dimensions are defined to handle your body, and therefore nothing else called a dimension can be. The definition of a line includes the word "endless", anything less is a line segment. If lines are endless, height is endless, width is endless, and breadth is endless, they leave no opportunity for another. If this universe is expanding limitlessly into the unseen universe, then there is no physical place allowed for another physical universe to exist in. If you want a third universe, you need to get another word because dimension is already taken; it has a definition. Go play with your goofy concepts somewhere else; we're talking about reality here. These dimensions and this world are rightly fit together with no need for fanciful machinations. A dimension has a strict definition, and it is used to confirm the structure of reality, whether this model is correct, or some other model is correct. Reality has a structure, just as surely as the apple falls to Earth, from a tree, under the influence of gravity. There are laws that govern physical interactions in our world, just as there are laws that govern the unseen universe. The same forces that God used to create the world are still in operation, just like gravity and the physical laws of the natural world.

Jesus said, "Truly, truly, I say to you, whoever hears my word and believes him who sent me has eternal life".

If we dare to believe the words of Jesus, as recorded in the bible, then we have everlasting life, now. We are citizens of both universes, now.

Jesus said to Peter, "I will give you the keys of the kingdom of heaven; whatever you bind on earth will be bound in heaven, and whatever you loose on earth will be loosed in heaven."

Believers have been given the keys to the kingdom, now. We have the power to move mountains, now. The gifts of the spirit can be a reality in our lives, now. That same reality is here, now. Did Jesus practice some kind of magic? No, he operated within the confines of the laws of the two universes in which he lived; by the principles that governed his reality; your reality; all reality. Did he cause natural things to be affected by supernatural laws? Yes, Jesus said to the fig tree, "May no one ever eat fruit from you again." and it withered from the roots. The only way that miracles were possible was for spiritual laws to be in effect, in the same place, at the same time, as the laws of physics and every property of the natural world. Jesus didn't possess some any kind of special magic or any other weirdness; he operated within the principles that governed his reality. When Peter marveled at the fig tree, this was Jesus' response...

"Have faith in God," Jesus answered. "Truly I tell you, if anyone says to this mountain, 'Go, throw yourself into the sea,' and does not doubt in their heart but believes that what they say will happen, it will be done for them. Therefore I tell you, whatever you ask for in prayer, believe that you have received it, and it will be yours. And when you stand praying, if you hold anything against anyone, forgive them, so that your Father in heaven may forgive you your sins."

After Jesus cast the demon out of the boy, the disciples came to Jesus in private and asked, "Why couldn't we drive it out?", and Jesus replied "Because you have so little faith. Truly I tell you, if you have faith as small as a mustard seed, you can say to this mountain, 'Move from here to there,' and it will move. Nothing will be impossible for you."

That same reality is here, now. Are you prepared to call fire down from heaven? Have you moved any mountain? The only thing standing between you and that reality is you and revelation. I have caused the natural world to be changed by supernatural means; my words and my faith have changed physical things in this physical world that were not accomplished within the laws of the natural world. Don't think you can hang out with me and learn to do this, I can't give you anything, I don't have anything to give you, and these are not parlor tricks; I can't just do whatever I want; I'm not a genie, I cannot grant you a wish. I repented for one of them because it was selfish, and I've since avoided exercising it for this reason. I will say this, it has never happened after conscious planning; it has only ever happened without any forethought, whatsoever, as though it weren't my thought, at all. It's hard to explain, they were my thoughts, but I experienced them with no forethought; I had no thought of speaking forth before I spoke; as if prompted, but some of them are not the things I would expect the spirit to prompt, and it leaves me without full understanding. This is my reality. If this is not your reality, it's because you do not have the faith that comes by believing. It's a heart issue. Do you need to be healed physically? It's a heart issue. Every issue is a heart issue. That one statement could change the entire medical profession if only everyone believed.

What's in your heart? How did it get there? How does it function? What does it do for you? How does your mind interact with it? What makes you who you are? The heart is a pump that delivers the history that you saved there, which carries emotional content prioritized according to the significance in which you deemed. Some things carry so little emotional content that you are indifferent, though you recognize that you remember or identify with the concept. seek (or not), receive (encounter), understand (or misjudge), believe (or reject), act (or not) rinse and repeat. Can you name anything that you do that did not enter into you by that path? You can receive something that you didn't seek, and perhaps the seeking wasn't a conscious effort on your part, but you were drawn towards something because of the desires of your heart, which were fashioned in this way. The vast preponderance of who you are, and what you've become, has come to pass through that process. If you do not seek, what makes you think that you deserve to receive?

Does God owe you a good life on a silver platter? Does anyone owe you anything on a silver platter? If I seem angered, please know that it is not anger directed towards you, for I am at war with the god of this world. I love each one of you. I can be harsh, but it is not personal. The bible says (about faith, hope, and love) that "of these the most important is love" and "without love you have nothing". It's clear that love is a central theme of the bible, but is it the most important theme? Is spiritual authority more important to God than love? I say yes. Nothing is more important to God than spiritual authority. Spiritual authority is the central theme of the bible; without it you cannot love the way God loves. How many times did Jesus put the Father in the preeminent position? Nearly everything that Jesus said was a reference to scripture or the Father's authority. There are volumes of material I could put here. I have also said it like this, nothing is more important than relationship; specifically your relationship to god must be your preeminent relationship, and all relationships in their proper order; god, family, church, friends. If you only got that one choice correct, it would lead directly down the path toward love. Love can be the most important fruit, but placing god in his rightful place in your life is the way to get to the fruit.

What is the meaning of life? We are to worship and glorify God, but each of us has a more personal meaning for our lives. God has a unique purpose for your life, and it is up to you to seek him to find out what that purpose is. That is the meaning of your life. Beyond God's purpose for your life, you are meaningless, at best, and an agent of the enemy in the extreme. There is a war going on. Whose side are you on? You're part of the problem, or you're part of the solution. You shall be judged for every word the leaves your mouth. How's that going to go for you?

74 Practically Speaking

If the model is true, then every field of study must align with it. And therefore it must be useful to be applied to the world in which we live. That is where you can find empirical evidence to support it. You see me using the model, in this book, primarily to interpret the bible because that is what's important to me, especially in this message I brought to you today, but I have used it to gain perspective on many fields of study. It's profitable in everything, as far as I can tell. Go for it, you can do this.

75 Challenge To the Whole World

I challenge the whole world at great risk of loss to myself, both on earth and after. I'm stand here, facing you, just like I did when I let that boy hit me on the playground, I'm prepared to suffer all insults for my beliefs. What the world does with this theory is not my responsibility. My job is simply to give it to the world.

What has been given to us belongs to us. The manner in which I decide to provide it to the world is up to me. God did not tell me to go into the entire world and proclaim this theory, he simply gave it to me, and I kept the secret until now, partially because I wanted him to know that I could keep a secret, in case he had other secrets to tell me. I see now, that I received this secret because I asked for it.

The answer to my request was so profound that I stopped asking, mostly because I understood that he would tell me, and my reality became that I should not ask for anything that I did not want the responsibility for. It is a very strange dynamic that permeated deeper into me.

Basically, from that time to this, I have not asked god for much for myself. In direct opposition to the fact that I know he wants to give me good gifts, just like he wants to give you good gifts, I was paralyzed by the reality that I would have to answer for everything that he gave me, and my cup already runneth over; I have already been given more than anyone, and therefore my responsibility to him was greater than anyone. It has silently dominated my life for nearly four decades.

I wrote the meat of this book in a couple of weeks and I must answer for every jot and tittle. My wife, who is a talented editor, isn't even going to get a chance to revise it. I know she would, but she does so much for me and the whole family and everyone around her, already, she doesn't need another task, and I am just going to let it rock. I have no doubt it is poorly written, but it flowed from me so easily that I have no intention in polishing it. Here it is, warts and all: the good, the bad, and the ugly; my life, filleted before you; the story of the theory is inseparable from my testimony because it was a gift from the giver, the Ancient of Days, the creator who created all things from nothing, with his words, which never return void. There, you have it in a nutshell; the model can even gain perspective on itself because everything fits in a single continuum of dimensions, which is useful for describing all of reality.

76 Notes to Self

Oh my, I just read a sixteen page note to myself, along with several additional documents, which I cannot share. Some things are simply too hot-off-the-press to be included in this book.

Wow, I just read another note that I wrote to myself on 10/29/2002, I can show you the timestamp on the file, and I'm going close this book by giving it to you here, unedited. It's not polished in any way, it's simply a note that I wrote to me, and it contains incomplete thoughts, which I'm not going to try to complete now. Every jot and tittle, exactly as I found it, just now, including a twenty-year period, which is completed in 2022, the same year in which this book will be released:

I've watched the world, and especially scientists, move toward supporting this revelation, little by little over the years.

The story includes the holy grail of science – the Grand Unified Theory that Einstein spent his life searching for. It has ramifications in every discipline and the world will reject it.

I chose this path more directly than I care to admit. I used the Word of God against myself, as a camouflage tactic to hide from the world. It led to double-mindedness. I seemed not to care while having ultimate concern.

I'm in travail and need a safe place to give birth, a safe place for what is being birthed, and safe for those exposed to the birth. I don't want to mislead anyone, or glorify myself.

Must subject my flesh to the fire of purification, and it doesn't want to go.

Sin of prayerlessness. Have to die to self and submit to His authority. Let Jesus be the Lord of my life.

God has taken some of it from me. I read some of what I had written, and had no recollection of the topic. It seemed as if it belonged to someone else.

I must subject the story to the Word, and the authority of the house, in order to burn away anything that doesn't hold up, before anyone is influenced by it. I might have to take it to the world and I only want them to see Jesus. Religious people will be especially susceptible to the beauty and simplicity of it. It's compelling, like what a false messiah would want to use, and I don't want to lead anyone astray.

Frankly, I feel as though I could use another 20 years of silence to prepare for what I might face, yet Jesus doesn't seem to be tarrying that long. Has He raised up another to do it? It will accelerate and intensify many things.

My model is such a simple model. Many will feel as if they already know it, yet it has never been expressed like this.

77 The End?

I told you a story that is wildly fanciful, and you have to decide whether it is true. Everything in here might not be correct, even by my own current view. I'm sure I could say some of these things better that I said them here. I rushed to publication without having the time to deal with everything included, some could be out of date from my own current belief, because the urgency of travail overtook me; the timing of this birth was not of my own choosing.

If I am allowed to live long enough, and have the burden to do so, you can expect more from me, including reproof of my own words, for which I'm responsible; god willing and the creek don't rise. Most of this content is decades old to me, little in here is hot-off-the-press in my life. I'm excited for what is coming. I'm completely captivated by god, and many things are settled in my heart. I wish the same for you, to be filled with his truth, which he desires to give you. May god, the creator of the physical universe, say good things about you, and to you, as you walk out his purpose for your life.

78 Be Saved and Avoid the Second Death

Romans 10:9

That if thou shalt confess with thy mouth the Lord Jesus, and shalt believe in thine heart that God hath raised him from the dead, thou shalt be saved. For with the heart man believeth unto righteousness; and with the mouth confession is made unto salvation.

79 Conclusion

This week began on Sunday, the first day of the week, on 12/12/2021. According to the astronomers, on that day, there were seven large objects lined up in a fairly straight line, visible without the need for magnification. They were five planets and two large asteroids. I think if you time it just right, you can also see an eighth, the moon. Do these represent the seven dimensions that you function in, plus eternity? It matters not, for I do not need signs in the sky to confirm anything, but it was a nice touch, as this week is special to me, for this week ends on Saturday, the seventh day of the week, on 12/18/2021, the first Sabbath that I shall observe.

Today, 12/17/2021, as I conclude the writing of the content in this book, I am convinced of all that I say, and before the light passes from this day, I shall pray the following prayer to commit myself to these things, which I have told you here, to keep the Sabbath as a sign between me and God. In a very short time, I recently experienced crisis in my life, both physically and spiritually, and it has caused the birth of this book, at this time in my life. Nothing remains to be done in this book, after today, except the publishing process to deliver this book to you.

80 Talk to God Through Your Intercessor, Jesus, to Settle These Matters in Your Life, With Me Now

My Father in heaven, the Holy One, Ancient of Days who sits on the thrown, forgive me, I have sinned and I know I need to be forgiven, and I trust that the sacrifice of your son Jesus on the cross is sufficient to pay the price for my sin. I need your word in me. Jesus, hear my prayer as you sit at his right hand, and reconcile me to your father in heaven. I give you my life that I might live with you forever. I accept your promises of mercy and sanctification by observing and doing your law as a sign to the world and I receive your seal on my forehead. I have no god but you. I reject all idols and images. I worship only you. I vow to keep your commandments. I shall not take your name in vain. I will remember the Sabbath and keep it holy. I shall honor my mother and my father. I shall not murder, I shall not commit adultery, I shall not steal, and I shall not tell lies about others. I do not want any one's spouse, or house, or anything that belongs to another. Father God, I trust that you will work your righteousness in me and I come to you through the blood of your son Jesus. Only you can work righteousness in me. I offer this sacrifice now, that you know, as a sign of my commitment to you, you know my struggle, you searched my heart. My sacrifice does not save me, only the blood of Jesus saves me. I seek your righteousness and this is the first of many offerings that I will bring to you as you work out my sanctification. In your timing, I ask for you to give me power to work miracles in your name, for your purpose on this earth, to bring your kingdom from heaven. I forgive others as I have been forgiven. Not my will, but your will, be done on earth, as it is in heaven, this day, in my life, in Jesus' name. Let it be so.

81 Go Forth and Be Baptized

Observe the Sabbath and receive the Seal of God on your forehead, to protect you from judgment on earth, while you live through the tribulation. Yes, I said through it. He's not coming near to pull you out, you don't get to leave here alive to escape the tribulation; it's coming and you're in it. His wrath is not designed for you, but this promise to be protected in it, can only be accepted by observing the Sabbath. God said so, and you must believe him, and act on that belief, now, before the door closes. Even so, come quickly Lord Jesus.

ABOUT THE AUTHOR

This is my first book, I scarcely read, much less write. I'm not an academic, but I understand that it's not the normal process for a book to be written in a week. It had to be released, that's all I can say. I didn't even receive a proof copy before I released it because I can't wait a month to see my work or know that I'm happy with the product. I'm not even going to read it, here it is. I fully expect to be less than thrilled with my work. It has to go, and it has to go now. This birth is happening on 1/1/2022.

Made in the USA
Monee, IL
14 January 2022

d3b7de10-605a-4ba9-800e-267fef443d77R01